HOW IT ENDS

ALSO BY CHRIS IMPEY

The Living Cosmos

HOW IT
ENDS

FROM YOU TO THE UNIVERSE

Chris Impey

W. W. NORTON & COMPANY

NEW YORK ■ LONDON

For information about permission to reproduce selections from this book,
write to Permissions, W. W. Norton & Company, Inc.,
500 Fifth Avenue, New York, NY 10110

For information about special discounts for bulk purchases, please contact
W. W. Norton Special Sales at specialsales@wwnorton.com or 800-233-4830

Manufacturing by RR Donnelley, Harrisonburg, VA
Book design by Ellen Cipriano
Production manager: Andrew Marasia

Library of Congress Cataloging-in-Publication Data

Impey, Chris.
   How it ends : from you to the universe / Chris Impey.—1st ed.
      p. cm.
   Includes bibliographical references and index.
   ISBN 978-0-393-06985-3 (hardcover)
1. End of the universe. I. Title.
   QB991.E53I47 2010
   523.1'9—dc22

                                                    2009047265

W. W. Norton & Company, Inc.
500 Fifth Avenue, New York, N.Y. 10110
www.wwnorton.com

W. W. Norton & Company Ltd.
Castle House, 75/76 Wells Street, London W1T 3QT

1 2 3 4 5 6 7 8 9 0

To K:
no beginning, no end

Contents

PREFACE   11

CHAPTER 1: ENDINGS ARE PERSONAL   17

That's All Folks   18
Everything Has Its Time   25
End of the Line   33

CHAPTER 2: ALL GOOD THINGS MUST PASS   41

The Reaper's Scythe   42
Beating the Odds   50
Dust to Dust   56

CHAPTER 3: THE FUTURE OF HUMANITY   65

The Fate of Species   67
Our Own Worst Enemy   79

CHAPTER 4: BEYOND NATURAL SELECTION   90

Endpoints of Evolution   91
Beyond Biology   102

CHAPTER 5: THE WEB OF LIFE   110

The Restless Earth   112
Gaia   124

CHAPTER 6: THREATS TO THE BIOSPHERE   134

A Hard Rain   136
Saving the Planet   142
Life Is Viral   149

CHAPTER 7: LIVING IN A SOLAR SYSTEM   157

Pale Blue Dots   158
Life Beyond Earth   166
Threats from Beyond   172

CHAPTER 8: THE SUN'S DEMISE   181

Living with a Star   183
Moving Off-Earth   192

CHAPTER 9: OUR GALACTIC HABITAT   203

Once and Future Stars   205
Mergers and Acquisitions   218

CHAPTER 10: AGING OF THE MILKY WAY   227

Fade to Black   229
Childhood's End   238

CHAPTER 11: HOW THE UNIVERSE ENDS   248

Something from Nothing   249
How It All Ends   261

CHAPTER 12: BEYOND ENDINGS   271

    Living in the Multiverse   273
    From Endings to Meaning   280

GLOSSARY   291

NOTES   301

READING LIST   325

CREDITS   329

INDEX   331

Preface

"The universe is made of stories, not of atoms," said poet and political activist Muriel Rukeyser. I agree. One of the greatest myths of science is that is consists of nothing more than dull, obdurate facts. The myth dissolves in the face of the powerful narrative that science has created to help us organize and understand the world. We have a story of how the universe grew from a jot of space-time to the splendor of 50 billion galaxies. We have a story of how a broth of molecules on the primeval Earth turned into flesh and blood. And we have a story of how one of the millions of species evolved to hold those 50 billion galaxies inside its head.

This is a book about endings. Science mostly answers the question of how things got to be the way they are. Yet if we stop at the present day, the job is only half done, as every good story needs an ending. Explanation is comforting but as the Danish cartoonist Storm P once said, "Prediction is very difficult, especially about the future." As a result, the material in this book is rooted in fact but it extends into conjecture. Scientists steer toward the boundary between what they know and what they don't know because that's where the excitement is. Despite the high proportion of speculation, I hope the reader finds the investment in fact more than trifling.

The material moves outward in scale from the human to the cosmic, and outward in time span from the familiar to the nearly eternal. In the first two chapters we confront our own deaths, and then consider the manner of our passing. The third chapter looks at threats to humanity and the fourth considers the likely fate of our species. As feisty apes with more piss and vinegar than wisdom we may not survive troubled adolescence, but visionaries are imagining ways we could transcend the limits of biology. The fifth chapter examines how we are webbed into the biosphere, and the next chapter looks at threats to the whole ecosystem. In all this, our atoms continue being part of the story.

In the second half of the narrative we move to the big picture of the future. There's no place like home for us to hop to if we mess up the planet, but beyond the Solar System there are likely to be millions of Earth clones. Going off-Earth may be the only way to keep our story going for billions of years. After considering habitable planets and the fate of the Sun, the narrative turns to our city—the Milky Way—and looks at the exotic fate of its stellar denizens. Finally, we project the fate of the universe and consider the possibility that this 14-billion-year saga might not be real, or the likelihood that it's just one of the stories that space and time have concocted.

They're esoteric, but the stories are about us. Even when considering our place among the galaxies, there are aspects of the universe that are conducive to our existence. The universe may not be mindful of us, but it turned the bed down and put a mint on the pillow like it knew we were coming. Time is the ruler for these stories. We follow it on scales from a heartbeat to the 10^{80} years it takes for the galaxy to dissipate. Physicist John Wheeler reminded us that we take it for granted when he said, "Time is what keeps things from happening all at once."

The writing is aimed at the general reader. I've tried to keep jargon to a minimum; essential terms are defined in a glossary. Technical details and asides are confined to the endnotes. The narrative is animated by vignettes at the beginning of each chapter, thumbnail sketches of top researchers, and even by some personal anecdotes, all of which serve as reminders that science is an essentially human activity, as complex and occasionally flawed as people themselves.

Everyone likes a good ending. But they're easier to relish when they're fictional, like the catharsis of a great movie or book, when the tension is resolved and all the loose ends are wrapped up neatly. This book is factual and it talks about the actual death of our planet, our star, our galaxy, *us*. It's not a cue to be glum, however, because the universe is filled with such magnificent possibility.

This project has taken me far beyond the bounds of my training and normal scholarship into chemistry, geology, biology, and sociology. I've benefited from conversations with Fred Adams, Nick Bostrom, Carol Cleland, Frank Drake, Carlos Frenk, Andrea Ghez, Richard Gott, David Grinspoon, Phil Hopkins, Lisa Kaltenegger, Michael Kearl, Ray Kurzweil, Chris McKay, Katy Pilachowski, Martin Rees, and colleagues across the University of Arizona. Any errors due to insecure grounding or over-reaching into alien fields are entirely mine.

I made heavy use of the Internet, so I thank Sergey Brin and Larry Page for keeping a billion Web pages indexed at my fingertips. If they could manage the trick of returning a search in the form of the answer to a question I asked, they'd really be onto something. I'm grateful to the Templeton Foundation for funding the project that brought many of the people in this book my way, and to NSF and NASA for funding my research on the science of endings large and small. I acknowledge the tranquil and reflective environment of the Aspen Center for Physics, where several of these chapters were written. Thanks to my agent Anna Ghosh for steering me through the shoals of the publishing world and finding good homes for my work. I acknowledge Angela von der Lippe at Norton for her expert guidance. I'm grateful to my friends, near and far, for their support and for rescuing me into the real world when I venture too far into the rabbit hole of writing.

CHRIS IMPEY
Tucson, Arizona
July 2009

HOW IT ENDS

Chapter 1

ENDINGS ARE PERSONAL

Patti Reynolds has been a hairsbreadth from death yet lived to tell the tale. In 1991, the young singer-songwriter was promoting a record with her husband when she was suddenly unable to speak. An MRI showed that she had an aneurism on her brain stem. It was already leaking, turning her into a ticking time bomb.

Within days a young surgeon in Arizona had chilled her body to 60 degrees and drained the blood out of her head, like oil from the sump of a car engine. This radical procedure—called cardiac standstill—was the prelude to snipping the aneurism and bringing her back from the edge of oblivion. The surgeon, Robert Spetzler, said at the time that Patti was as deeply comatose as anyone could be and still be considered alive. During the operation Patti's eyes were taped and she had molded speakers in her ears.

And then, as she puts it, she popped out of her head. She was high up in the room and looked down to see 20 people clustered around her at the operating table. She heard the sound of a dentist's drill, and the surgeon and a nurse speaking. She noticed a tunnel and a bright light and she talked to her dead grandmother and uncle. As the doctors restarted her heart, she heard the Eagles song "Hotel California" playing in the operating theater and she relished the irony of the line "You can check out anytime you like, but you can never leave."

Patti assumed she had been hallucinating, but all of the details were subsequently confirmed by witnesses and hospital records. How could she have seen these things as she laid flat-lined on a gurney? Spetzler said he had absolutely no scientific explanation for what happened. To skeptics, the awareness of a near-death experience and the associated imagery that has almost become a cliché—tunnels and white lights—are symptoms of altered brain function caused by extreme oxygen starvation.

But the growing anecdotal evidence of near-death experience includes examples where conscious memories are formed when electrical activity has been reduced to an undetectable level, raising the question, when the brain shuts down, where does the mind go?

That's All Folks

Facing the Inevitable

Some mornings when I'm drinking coffee, early morning light glints off a gold mask on my living room wall. The face is in repose, the nose is straight and noble, and the eyes are half-lidded. It's Agamemnon, the commander of the Greek army during the Trojan War. The original was discovered by Heinrich Schliemann in 1876 in a tomb at Mycenae, and it helped persuade scholars that the epic stories of Homer represented real events and real people over 3000 years ago.

Gazing at the serenity of Agamemnon, the inevitability of death seems almost palatable. How noble it would be to die as a king and a warrior. Agamemnon was flawed, as were all the Greek heroes. He sacrificed his daughter Iphigenia before setting sail for Troy, and he quarreled with Achilles incessantly, but eventually he led the Greeks to victory and performed many heroic feats on the battlefield. He returned home victorious with a concubine, Cassandra, seized as a spoil of war. His wife Clytemnestra, meanwhile, had taken her own lover. Agamemnon was treacherously slain at a banquet in his honor.

We're all going to die. It's natural to harbor hopes of an ending that's epic or tragic. Yet while the mask of Agamemnon shines brightly, it's an illusion. The graves at Mycenae were subsequently shown to have been constructed 300 years earlier than the conjectured date of the Trojan War. The gold mask is fine enough that it may be the likeness of a nobleman or a king but there's no evidence that it represents the historical Agamemnon. There's even strong suspicion that Schliemann salted the grave site with the mask to burnish his reputation.[1] Various other accounts deconstruct the death scene. The commentator Pindar wrote that Agamemnon was in his bath and was slain by his wife after she tossed a blanket over him to prevent any resistance.

So it is with our own deaths. Few of us will meet an end that is quick and larger than life, like James Dean, who died in a head-on collision at age 24, or Christa McAuliffe, the schoolteacher who died when the space shuttle *Challenger* exploded 73 seconds after launch. Our great fear is that our bodies will painfully wear out and fail us, or our brains will slowly unravel. Unfortunately life rarely mimics art. If life is like a movie, it doesn't end with a climactic scene; more often the celluloid gets grainy and frayed and the actors forget their lines. If life is like a concert, it doesn't end in a crescendo; more often the

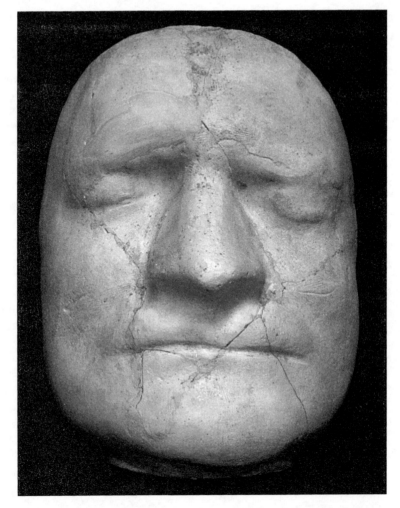

Figure 1.1. The death masks of antiquity, such as those of Agamemnon and Tutankhamen, were not sculpted from the deceased person. By the time of Isaac Newton, however, true likenesses of famous people were being made from casts of wax or plaster. This is a copy of an original mask of Newton held by the Royal Society in London.

instruments go out of tune, the musicians wander off, and the music peters out (Figure 1.1).

To start this exploration of endings, we must acknowledge and move beyond the obstacle of our own demise. It seems so unfair. All of our dreams and aspirations will be truncated, perhaps by a wayward bus at age 40 or maybe by a rogue tumor at age 80. Human achievement is an occasionally noble, but ultimately futile, effort in the face of an inevitable outcome. The only uncertainty, except in the case of suicide, is the manner of our departure.

The Art of Death

Facing the unthinkable—what Rabelais called the "vast perhaps"—has caused humans many sleepless nights but it has also spurred some of their greatest creativity. From Lear's lament over his dead daughter Cordelia to Thomas Pynchon's epic ruminations in *Gravity's Rainbow*, death has exercised the minds and imaginations of writers. It swirls like the pumpernickel in marble rye through much of the best fiction, poetry, and theater of the past century, sometimes symbolic and metaphorical, sometimes cool and ironic, sometimes intense and confessional.

Literature often employs death as a vehicle to comment and reflect on life and the nature of existence. Thornton Wilder's play *Our Town* uses the evanescence of life to gently remind us of the daily wonders that surround us. In Henry Thoreau's *Walden*, a similar perspective derives from communing with nature. Leo Tolstoy's protagonist Ivan Ilyich faces an incurable illness and comes to terms with his predicament by asking for help. In the novels of Franz Kafka and D. H. Lawrence, and the plays of Eugene O'Neill and Tennessee Williams, the concern with mortality is almost obsessive. Poetry goes even further, stripping the subject matter down to its raw essence. The acute sensibility of Sylvia Plath and Anne Sexton is undoubtedly heightened by our knowledge that they took their own lives.

Mortality is both a truism and a taboo. Western countries cultivate a careful posture of denial about actual death, preferring the stylized or archetypal representations in the mass media. Yet for most of human history, and in poor countries to the present day, life and death have been seen as natural processes, part of a continuum. Death was not held at any distance; it was vital and real. Lest anyone forget, there were constant reminders. Victorious generals in Imperial Rome were accompanied in victory parades by slaves who intoned, "Remember, you are mortal," and by the Middle Ages these reminders had a moral overlay, as in the example of ivory statues with a beautiful woman on one side and a rotting corpse on the other.

Modern Western culture manifests an unhealthy schizophrenia about mortality. On the one hand it fuels our obsession with youth and life's ephemera, while on the other hand it pounds us with images of real and synthesized violence and mayhem. There's irony in that, because graphic representations of death do nothing to lessen its mystery; we lose our innocence and gain nothing in return. Using Walt Disney as a lens, the death of Bambi's mother was only hinted at, but 50 years later in *The Lion King*, Mufasa plunges to his death and little is left to the imagination. The culmination of this progression is the first-person shooter video game, where realism is paramount and the players are confronted with the dreck of red electron splatter.[2]

In art, staring at death with a gimlet eye takes bravery; too often the sensibility is morbid or maudlin. The delicacy of classical metaphor has been superseded by realism. Damien Hirst's diamond-encrusted skull sold at auction for a staggering $100 million, and Günther von Hagen's traveling collection of dissected cadavers has been seen by 25 million people worldwide. We shouldn't be surprised that Gregor Schneider is looking for volunteers willing to die in an art gallery.

When it all gets too heavy, we can retreat into quips and aphorisms, like Woody Allen's one-liners, or this example from Art Buchwald, as he suffered through terminal kidney failure: "Dying is easy. Parking is hard." In their cartoon world of aching chasms and obdurate rock, the coyote and the roadrunner are endlessly reborn (though perhaps each time it's a different coyote and a different roadrunner). But before the black iris closes shut the rascally rabbit pops out to tell us not to take it too seriously.

Talking About Endings

Is there a perfect ending? At its best, death might be grimly fitting, such as IRA car bombers or Hizbullah suicide bombers who died at their own hands. Or numerically apropros, such as chess whiz Bobby Fischer dying at age 64. Or tragic-comedic, such as the actor Harold Norman, who accidentally died in 1947 during an overly energetic sword fight in the ultimate scene from *Macbeth*. Or poetically fitting, such as Attila the Hun dying of a nosebleed on his wedding night. Or perverse, which would be a toss-up between running guru Jim Fixx dying while jogging and Irving Rodale, the founding father of the organic food movement, dropping dead from a heart attack while he was being interviewed on *The Dick Cavett Show*.

Perfection need not be epic. A good case can be made for Don Duane, who checked out at a Ravenna, Michigan, bowling alley in 2008. Don had been a member of a local team for 45 years, but on this October night, after rolling his first perfect 300 game, he high-fived his teammates and collapsed from a massive heart attack.

The people who choose to talk about death (apart from doctors and morticians and actuaries) are not what you might expect. To fend off the casually morbid, Mike Kearl avers that his expertise is thanatology so that only those savvy in Greek will realize he's talking about death. Kearl is a professor emeritus at Trinity College and he has authored dozens of papers on death and dying, as well as contributing to three encyclopedias on the subject and maintaining a massive Web site on endings that he uses extensively in his teaching. This level-headed approach is welcomed because our perception of a threat isn't always proportional to the actual health risk (Figure 1.2).

Kearl claims that he's not morbid, though he does admit to an Addams family motif in his house, and he's been given a lot of tombstones as Christmas gifts, which pile up in his backyard. (He's dropped hints to his friends about a gift he'd prefer—a pen with Lincoln's DNA in it, for just $1600.) He thinks that the Western cultural aversion to death is very strong, and says Americans are "a great death-denying culture." This even extends to the people who must deal with death; only 5 of the 126 medical schools in the U.S. have required courses on death and dying. The denial couples to a growing bullishness on the beyond. The fraction of adults believing in life after death has grown from 15 percent to over 40 percent in the past 30 years.[3]

Author Michael Largo has delved even more deeply in the subject of death, to the degree where he says his friends worry about him. He rebuffs them with a quote from the Dalai Lama: "If you are not aware

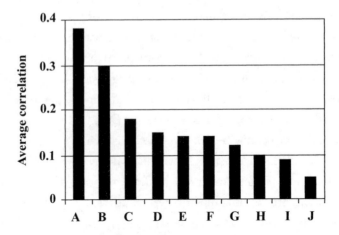

A. Smoking and lung cancer
B. Media violence and aggression
C. Condom use and sexually transmitted HIV
D. Passive smoking and lung cancer at work
E. Exposure to lead and IQ scores in children
F. Nicotine patch and smoking cessation
G. Calcium intake and bone mass
H. Homework and academic achievement
I. Exposure to asbestos and laryngeal cancer
J. Self-examination and extent of breast cancer

Figure 1.2. The correlation between behaviors and public heath threats in the United States, based on a review of 50 years of research by L. Rowell Huesmann from the University of Michigan. The category "media violence" includes TV, movies, and video games.

of death, you will fail to take advantage of this special human life." The son of a New York City homicide cop, Largo says his father would "point out the Empire State Building and say it's 1200 feet tall and 61 people have jumped off it. He'd always add some death fact. That was the only thing I listened to. I realized there were so many ways to die."[4] Largo turned this interest into murder mysteries and a series of nonfiction books, earning him an informal crown as the "king of kaput."

Largo spent 10 years gathering statistics and anecdotes for his bleak-sounding compendium *Final Exits*, yet the book's a surprisingly jaunty read. He notes that death is getting more interesting. In 1700, there were less than a hundred causes of death listed on death certificates; now there are over 3000. Even traditionally exciting and dangerous activities have spawned new types of exit—think of base jumping, which manages to combine the hazards of climbing and skydiving.

He guarantees that readers of his book will gain a minimum of two extra years of life, though a little thought shows that it would be very hard to verify his guarantee. Largo can reassure, or startle, with the odds of dying in a revolving door, or on a waterbed, or while praying, or by laughing or hiccupping. He says, "Death is a moody mother" and when pressed for a particular example of the perfidy of fate, mentions the man who survived the collapse of the Twin Towers only to die 30 months later in the Staten Island ferry crash that killed 13.

The classical standard for bravery—confronting a noble death in the heat of battle—was established by Plato in *The Republic* and Aristotle in *Nicomachean Ethics*. This is far too narrow; there's bravery aplenty in the myriad ways humans acknowledge the enigma of death. These accommodations are particularly interesting when they occur without the succor of religious faith.

Consider this account by the biographer James Boswell, who was a believer, from the deathbed of the noted philosopher David Hume, who was an atheist: "I asked him if the thought of annihilation ever gave him any uneasiness. He said no, not in the least, not more than the thought that he had not existed, as Lucretius had observed."[5] According to his wife Ann Druyan, as writer and astronomer Carl Sagan lay dying from leukemia, "there was no deathbed conversion, no last minute refuge taken in a comforting vision of a heaven or an afterlife. For Carl, what mattered most was what was true, not just what would make us feel better. Even at this moment when anyone would be forgiven for turning away from the reality of our situation, Carl was unflinching."[6]

Everything Has Its Time

A Little Perspective

Let's get a little perspective on our mortal coil. Average life expectancy at birth in a developed Western country is about 80 years or 2.5 billion seconds. Viewed in terms of the ticking of a clock or the steady sweep of its second hand, that seems like an eternity. Our life spans last 1 percent of recorded history and less than 0.1 percent of the history of modern *Homo sapiens* so far. Recent population growth creates huge compression in this history; most of the people who have ever lived are alive now.

However, humans are late and recent arrivals on an ancient planet. It was 400 million years ago that the first animals crawled from the seas and began to live on land. The origin of life itself was much earlier still, about 4 billion years ago, at a time when Earth's atmosphere had no oxygen and its surface was being pounded by debris left over from the formation of the Solar System. The Earth-Sun system is in turn a relative newcomer in a vast universe that houses billions of trillions of stars and their attendant planets. Observations of tepid microwaves left over from the hot big bang creation establish the current age as 13.7 billion years.

To get a grip on this vast track of time, let's shrink or accelerate the history of the universe by a factor of 13.7 billion. Imagine it's now the stroke of midnight on New Year's Eve and the big bang took place at the same time last year. Planet Earth forms in mid-September, cells with nuclei first appear in mid-November, and animals begin colonizing the land on December 21. The first humans evolve just an hour and a half before midnight on December 31. The glory of the Renaissance, the Agricultural and Industrial revolutions, the Space Age, and the rise of computer technology all fit inside the last second of this cosmic year.

A human life in this scale model is little more than a tenth of a second. If the universe has existed for a year, all our personal hopes, dreams, and ambitions are squashed into the blink of an eye. As the film *This Is Spinal Tap* reminded us, it's possible to have too much perspective.

Older Than Dirt

What are the oldest living creatures? Bacteria can potentially survive forever, but only as a colony; stem cells and gametes are in the same category. Protozoans are animals, but they reproduce by division from a single parent. We concentrate on individuals that result from sexual reproduction, and pay less attention to groups of genetically identical organisms. Soil can form in as little as a few thousand years, so some of these creatures are indeed older than dirt.

The oldest living plant is a bristlecone pine tree nicknamed "Methuselah," whose exact location in the Mohave Desert is kept secret by the U.S. Forest Service to protect it from vandalism. In 2008 Methuselah celebrated (but not very boisterously) its 4840th birthday. Even older plants are found in clonal colonies where new growth emerges from a long-lived root system, though the original old growth has long since disappeared.[7] A creosote bush also in the Mohave Desert has lived for over 11,000 years, starting small and growing outward in a circle. In Sweden, there's a spruce tree whose root system first got established at the end of the last Ice Age, 9500 years ago. One unique shrub that spreads over a kilometer of Tasmanian rainforest is 43,600 years old, and an Aspen grove in central Utah is estimated to be 80,000 years old. But these land plants are eclipsed by a colony of sea grass that surrounds the Mediterranean island of Ibiza; some of its strands are 100,000 years old.

Among sea creatures, if you want to live to be old, it's best to be cold. A slow-growing Antarctic sponge was estimated to be 1550 years old. In 2007, a 405-year-old clam was retrieved from the Arctic Ocean. Its age was measured using the shell growth layers that are laid down in each warm season, analogous to tree rings. The clam was a youngster when Shakespeare penned these words in *The Tempest*, "Full fathom five my father lies; of his bones are coral made; those are pearls that were his eyes; nothing of him that doth fade." It escaped capture by Norse travelers to America, whose settlements predate the Pilgrims' arrival at Plymouth. Thankfully, the researchers who found it resisted turning it into chowder.

Fish can also live to be very old, but they move around freely so it's hard to keep track of them. Hanako, however, spent his life confined to a pond in Nagoya, Japan. He was a koi fish, or a carp, who died in 1977 at the age of 226. The president of a women's college where the koi lived told how Hanako would respond to his name, welcome a pat on the head, or even tolerate being pulled out of the water for a hug. Rockfish are more elusive—their bulbous eyes, fleshy lips, and jowls make them fish that only a mother could love—but they can also live for over 200 years. Even unassuming goldfish can live into their 40s.

Tortoises typically hold the record for the oldest living animal. Probably the most famous was a Galapagos land tortoise who spent most of his life at the Australia Zoo in Brisbane. Harry was five years old, and the size of a dinner plate, when Charles Darwin brought him to England on the HMS *Beagle*. Harry was miserable in the English climate, reduced to a state of almost permanent hibernation, so he was sent on to Australia. Females were introduced to his enclosure for a century with no result, but nobody had taken the trouble (or been able) to flip the 136-kilogram tortoise over and check the gender; Harry was a she. Harriet retired from giving children rides on her back at age 150, and enjoyed a diet of parsley, endives, squash, bok choy, and hibiscus, plus a daily bath to remove bird droppings, until her death in 2006 at the age of 176.

A radiated tortoise from Madagascar lived even longer, from the year after American independence to the year after the Beatles' invasion of America. Tu'i Malila was presented to the Tongan royal family in 1777 by Captain James Cook. The Tongans clearly were impressed by Tu'i Malila; its cage was the first stop on Queen Elizabeth II's tour of the island in 1953. Another tortoise belonging to General Robert Clive of the East India Company may have lived as long as 250 years but is the subject of an age dispute. Nobody's in a hurry to sort it out.

Living Long, Living Large

People who pore over the records of zoos and aquariums all over the world have long known of a relationship between the size of an animal and its maximum age. In round numbers in years, among mammals, elephants live to 70, lions to 30, wolves to 15, rabbits to 10, and mice to 5. Among reptiles, there's a maximum of 150 for giant tortoises, 70 for alligators, and 20 for cottonmouth snakes. With birds, we see up to 120 for turkey buzzards, 80 for parrots, 20 for canaries, and 10 for the hummingbird. The same pattern is seen among fish and amphibians.[8]

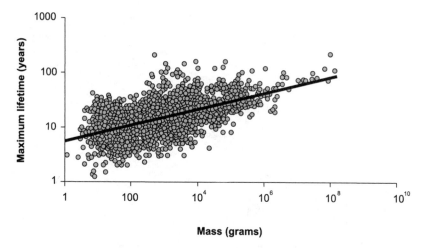

Figure 1.3. There's a correlation between maximum life span (in years) and adult body mass (in grams), here shown for 1700 species from the Animal Ageing and Longevity Database, produced by João Pedro de Magalhães of the University of Liverpool. The correlation is relatively flat—a factor of 10 gain in longevity occurs over a factor of 3 million gain in mass—and the scatter is large.

Throw all these data in the pot and there's a linear correlation between adult body size and maximum life span (Figure 1.3). However, there's a big scatter and there are some notable exceptions: Bats and eels, for example, live longer than expected for their size. One explanation is ecological constraints; smaller animals might be more susceptible to predators so on average they would die quicker. Another well-known but controversial result is the correlation between metabolic rate and body mass (Figure 1.4). Since size or mass correlates with longevity, that means metabolism is related to life span. Perhaps reptiles and amphibians live longer than similar-sized mammals because they're cold-blooded and have slower metabolisms. Similarly, mice live faster and so die quicker than people or elephants. This sounds appealing: each body gets an allotment of a billion heartbeats and they can be used quickly or slowly.

Unfortunately, the situation is more complicated than that. After the effect of body size is removed, there's a correlation between brain weight and maximum life span for primates. That doesn't mean that causes of aging are located in the brain; bigger brains may be helpful in helping animals escape predators. Within a species, the correlation

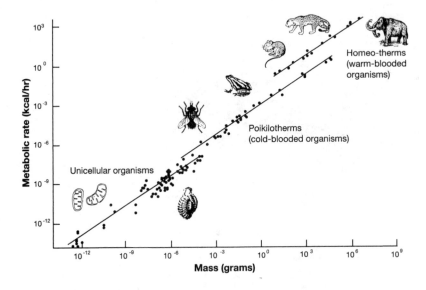

Figure 1.4. Many studies have found a correlation between metabolic rate, here measured in thousands of calories per hour, and mass or size. The relationship applies over an extraordinary range of a billion billion in mass, from single-cell organisms to large mammals. Because it couples to the life span–mass correlation, the slower metabolisms are associated with the longer lives. Researchers disagree on whether this is because of genetic factors or the influence of the environment.

between size and longevity can go in the opposite direction. There's evidence that short people live longer than tall people and pet owners know that small dogs live longer on average than large dogs.

Many of these subtle relationships have been studied and documented by João Pedro de Magalhães, a young researcher who leads a group on genomic approaches to aging at the University of Liverpool. He makes available on his Web sites databases on animal ages and longevity and on genes implicated in human aging. Freely admitting to be terrified of death, he wants to conquer aging "because it is and will be the major cause of suffering and death among the ones I love." He leavens the existential angst by performing music and stand-up comedy. He gives a due nod to his Sisyphean task with a quote from Churchill—"Success consists of going from failure to failure without loss of enthusiasm"—and another from science fiction writer Robert Heinlein—"Always listen to experts. They'll tell you what can't be done and why. Then do it."

What about exercise, which sounds bad because we spend time using up heartbeats more quickly? The math is easy. A 40-year-old woman in average health has a resting heart rate of 75 beats per minute, or 750,000 a week. Suppose she exercises five hours a week, during which her rate rises to 120, and in doing so she lowers her resting heart rate to 65 beats per minute, which is typical of someone in good physical condition. She has burned an "extra" 13,500 heartbeats, but she has "saved" 100,000. Plus, the heart has become a stronger machine as a result. It's not even close—never use metabolic life span arguments to mothball the Stairmaster.

Lest we get too obsessive about our heartbeats and life span, spare a thought for the poor mayfly. These insects belong to a group with the delightful name Ephemeroptera, meaning short-lived flyer. The female of one species that is crucial for the health of streams and lakes lives dormant underwater for a year in its nymph form, then it emerges for five minutes as a flying adult. During this brief life it must find a mate, copulate, and lay its eggs back into the water.

The Greatest of the Apes

The Great Apes is the category of primate that includes humans. Our genetic "cousins" rarely live longer than 30 or 40 years in the wild but can live longer in captivity. A lowland gorilla called Jenny died in 2008, four months after tucking into a four-layer frozen fruitcake and banana leaf-wrapped treats for her 55th birthday.

The oldest chimp is also the most celebrated chimp: Cheeta. Turning 76 in 2008, Cheeta is in the *Guinness Book of World Records* as the oldest nonhuman primate. Cheeta appeared in 12 Tarzan movies with Johnny Weismuller and Lex Barker from 1934 to 1949, he played Ramona in *Bela Lugosi Meets a Brooklyn Gorilla*, and his final role was as Chee-Chee alongside Rex Harrison in *Doctor Doolittle*. He has since retired to Palm Springs where he paints, leafs through books, plays the piano, and watches old movies with his grandson. Sales of his artwork have raised $10,000 for charity and Jane Goodall's reserve at Gombe.

His ghost-written autobiography, *Me Cheeta*, was published in 2008. Despite his fame, Cheeta has suffered the ignominy of losing out to Tinkerbell for a star on the Hollywood Walk of Fame.

And what of the greatest of the apes? Living to be 100 is rare but no longer remarkable. Worldwide, there are half a million centenarians, with about 60,000 in the United States and 30,000 in Japan (Figure 1.5). In the United States, they get a form letter from the president and a nod on NBC's *The Today Show*.[9] All U.K. and Commonwealth citizens get royal greetings on the big day and on each birthday they survive beyond 100. My Scottish grandmother and great-grandmother proudly displayed on the mantle their telegrams from Queen Elizabeth. The luck of the Irish nets centenarians there a "bounty" of 2540 euros. Past 100, people drop like flies. Only 1 in 1000 centenarians lives to 110, called a supercentenarian, and only 1 in 50 supercentenarians makes it to 115.

Discounting tales of 150-year-old wizened denizens of the Caucasus who thrive on brutal winters, sour cheese, and goat's blood, the oldest person who ever lived was Jeanne Louise Calment. She was born in Arles in 1875 and died in 1977 at the age of 122. At age 13, she met the painter Vincent van Gogh, describing him as "dirty, badly-dressed and disagreeable." Calment took up fencing at 85 and rode a bicycle until she was 100. At age 90, she sold her apartment to a lawyer with something called a reverse mortgage, where he agreed to pay her a fixed sum until she died. As a result he paid her more than twice the value of the apartment. A sharp lady in more ways than one, she once said, "I've only got one wrinkle and I'm sitting on it."

Naturally we might pore over the runes of Calment's story to see what was so special. Apart from longevity in her close family, the signals are mixed. It definitely helped that she married a wealthy store owner and led a life of leisure. (Her husband was not so lucky, dying 55 years before she did after eating spoiled cherries in his dessert.) She put olive oil on all her food and rubbed it into her skin, drank port wine, and ate two pounds of chocolate a week. She also smoked until she was 117, stopping only when her sight was so bad she couldn't light the cigarette.

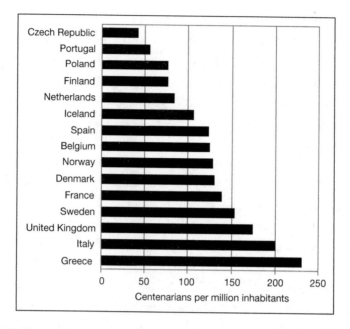

Figure 1.5. The number of centenarians in selected European countries in 2007 per million inhabitants. There's a variation of a factor of 5, not obviously explained by affluence or GDP per capita or lifestyle aspects such as the "Mediterranean diet." Genetics must play a strong role in these variations.

In the end, her "container" failed her, as it will one day fail for us all. The thought that the richness of life is followed by the absoluteness of nothing is disconcerting, to say the least. Jean Louise Calment's atoms have scattered into new configurations. Memories of her will persist for a while in those who knew her. Evidence of her will endure far longer in the form of the printed word in this and other books. Yet words are pale shadows of the person that was Jean Louise Calment. And so we inspect—gingerly and with some trepidation—the finality of death.

End of the Line

A Knotty Philosophical Issue

For thousands of years, human cultures have embraced the idea of a personal afterlife, in which some sense of the individual persists after death, perhaps to remain in the immaterial realm and perhaps to be reborn in another material vessel. The nature of this persistent state varies. To ancient Greeks, it was the bleak landscape of Hades. The monotheistic religions postulate a complex layering from paradise to damnation. In some Eastern religions, the soul can transmigrate into animals, plants, and even rocks.

The afterlife has a moral underpinning in most religions; it's deferred compensation (or retribution) for a person's conduct in their biological life. The possibility of life after death centers on one knotty issue: The problem of identity.

Plutarch wrote about a ship belonging to Theseus, which was exhibited in third-century BC Greece long after his time. In his words, "The ship wherein Theseus and the youth of Athens returned had thirty oars, and it was preserved by the Athenians down even to the time of Demetrius Phalereus, for they took away the old planks as they decayed, putting in new and stronger timber in their place. This ship became a standing example among the philosophers for the logical question of things that grow; one side holding that the ship remained the same, and the other contending that it was not the same."

The Athenians replaced each plank in the original ship of Theseus as it aged or decayed, keeping it in excellent condition. Many years later did they have the same ship that belonged to Theseus? A similar story is told of George Washington's axe, still claimed to be a relic of the first president's even though the head and handle have been replaced several times. Then there's the Puerto Rican boy band called Menudo, which has been making similar music for 30 years, despite a continual churn in the membership of the band.

Change and persistence apply not just to ships but also to human vessels. You're not the same person you were 10 years ago; that's the average age of your cells. The durability of cells depends on their role in the body's battlefield. Stomach cells last just five days, red blood cells are worn out after three months and 1000 miles of travel, liver cells last under a year, and even the skeleton is renewed every decade by an army of bone-dissolving and bone-rebuilding cells. Only lens cells inside the eye and the neurons of the cerebral cortex seem be inert and share the age of the person.

Let's look at two variations to the story of Theseus's ship that sharpen the conundrum of identity, courtesy of Marc Cohen at the University of Washington. First, suppose Theseus set to sea with all the replacement parts for his ship in the hold. While at sea he completely replaces the old parts with the new ones. Is the ship he returns in the same one he departed in? If not, how many of the original parts would he need to have left in for it to be the same?

In the second version, Theseus is followed by a scavenger ship that collects the pieces as Theseus discards them and builds his original ship anew. The two ships dock side by side. Which is Theseus's ship in this case? If we say his ship is the one that was reassembled from the discarded bits, then it's not the same ship he left on, so he somehow "changed ships." But if we say his ship is the one he ended his voyage on, then no part of it is in common with the ship he left on, while the ship sitting next to him after he docks has every part in common with the ship he left on, yet it's a different ship!

We can imagine philosophers whiling away happy hours with this stuff, but what has it got to do with the "great beyond"? If death really is the end—a closed door on all memory, experience, and personal identity—there's nothing to discuss. But if the door is open to transcendence of the biological organism, it's fair to ask what that might be like.

The Dilemma of Dualism

Probably not like this: In the 1990 movie *Ghost*, Patrick Swayze and Demi Moore play a happily married couple whose world is shattered when Swayze is murdered one night by a thief. Swayze becomes a ghost, trapped between worlds. He realizes his wife is under threat from a former co-worker, but he is powerless to intervene until he enlists the help of a con artist and medium, played by Whoopi Goldberg. In the movie, dead people rise from their corpses and have a diaphanous existence. They pass through people, walls, and other solid objects. They look human but are invisible to all but a few special people.

Hollywood is placing a safe bet with a film like this. According to an AP poll in 2007, one in four Americans has seen a ghost or felt a spiritual presence. Belief in angels and demons runs even higher, about 70 percent. Three in four Americans believe in an afterlife, probably unsurprising in a religious country where 90 percent believe in God.[10]

Belief in an afterlife attaches directly to a philosophical position called dualism. Dualism holds that humans consist of a material body and an immaterial soul. Similarly, dualism applied to cognition states that we have a physical brain and a nonphysical mind; they're not the same thing. These ideas date back to Plato and Aristotle, and were codified by the seventeenth-century philosopher René Descartes. He decided the mind and body were completely different, because he could doubt whether he had a body but he couldn't doubt whether he had a mind.[11] As he wrote in his book *Meditations on First Philosophy*: "I have a clear and distinct idea of myself as a thinking, non-extended thing, and a clear and distinct idea of body as an extended and non-thinking thing. Whatever I can conceive of clearly and distinctly, so God can create. And the mind, a thinking thing, can exist apart from its extended body. Therefore, the mind is a substance distinct from the body, a substance whose essence is thought."

Dualism *seems* reasonable because mental states appear so different from physical states. There's a tree in my backyard that anyone can see and although none of us can observe an electron directly, anyone can detect it in the same way by using instruments. On the other hand, my mental states are privileged; no one can share them. That's why I can be skeptical of other people's minds, but not my own. It's easy to describe the physical states that correspond to a burned finger or the blue sky or a musical sound. However, the experience of feeling pain or seeing blue or hearing music is far harder to define. Dualists argue that the subjective aspects of mental events—called qualia—cannot be reduced to anything physical.

Modern science has made it hard to defend dualism since there are so many couplings and causal connections between physical and

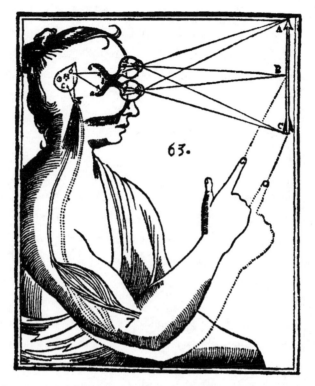

Figure 1.6. Descartes identified the mind with consciousness and self-awareness and the brain with intelligence. This illustration from his "Metaphysical Meditations" explains the function of the pineal gland, which he believed was an intermediary between the sensory inputs and the immaterial spirit. He called it the "seat of the soul."

mental events. I walk to the coffeepot because I want coffee, then I pull my hand away because I feel a burning sensation, and that sensation is caused by the burner on the stove being hot. Neuroscientists can now correlate mental states with localized electrochemical activity in the brain. Cartesian dualism posits an immaterial realm that's completely independent of the physical realm (Figure 1.6). Yet thought processes and attributes that form the essence of a soul manifest as actions in the physical world. We start life as a single cell made of matter and our development can be explained without invoking any nonmaterial mechanisms.[12]

Dualists do not identify a person with his or her body or brain, so the door is open for belief in an afterlife. In a supernatural and immaterial realm, anything's possible. In the story of Theseus, it's as if there's an essential "shipness" that's independent of all the wood. We cannot rule out disembodied survival on logical grounds, but have no idea how to describe what it would be like. Probably not like a movie.

The Materialist Quandary

Bodily resurrection is the standard view of the afterlife in the major theistic religions: Judaism, Christianity, and Islam. Given materialism, is resurrection possible? Because the atoms of any living organism are in constant flux, resurrection presents a huge challenge. How can the gap in time and space between the body that perishes and the resurrected body be bridged, such that the resurrected person is identical to the person who died?

The gap has to be bridged, because it's a central tenet of Christianity that the crucified Jesus of Nazareth had been, in a literal, bodily sense, raised from the dead, and his resurrection was the means by which the general resurrection of the dead would be accomplished. Not only that, but resurrection is not just restoration of the state of a person before death, as in the story of Lazarus, it's a doorway to a new kind of life. After resurrection, Jesus is still an organism, and he still has a body made of up quarks, down quarks, and electrons, but they have been reorganized into the perfectly functioning physical form that its perfect designer intended.

Suppose God wants to raise Socrates from the dead. How will He do this? To the dualist, there's no problem bringing Socrates back because, in a sense, he never left. All that's needed is a fresh body to house his soul. The materialist, however, knows that two millennia are enough to spread Socrates' atoms through the biosphere. This is a difficult job of reassembly, even given omnipotence. And if God can reassemble the atoms that comprised Socrates at the moment of his death in 399 BC, He can also assemble the atoms that comprised Socrates when he was much younger, in 440 BC. In fact, because there's no overlap in the sets of atoms, he could assemble both side by side. Socrates is neither or both, and because not both, neither. We're back to Theseus and his ships.

Peter van Inwagen, who's a philosopher and Christian scholar at Notre Dame, has taken these arguments even further. He suggests that God could, at the moment of death, remove the corpse and replace it with a perfect replica or simulacrum, which is what burns or rots. Or perhaps He removes for safekeeping a vital part of the core person. Continuity is maintained. When it's time for the resurrection, God restores life to the body in question. Identity-preserving materialist resurrection may not be impossible, but great ingenuity is required to pull it off.

Nontheistic religions that affirm an afterlife have a different quandary. Buddhism, Jainism, and some Hindu sects believe that reincarnation is driven by karma: good or bad deeds in this or previous lives. But this creates a karma "management" problem, because your genes and the family environment you find yourself in must correspond to the moral worth of your past deeds. The laws of nature are subtle and complex but they are also impersonal; physical situations pay no heed to moral considerations. So the karmic moral order of Indian traditions must be radically different from the order that governs the physical world. Yet the two are supposed to be intimately related.

Beyond the Body

Having bumped into the boundary between faith and reason, the only place to turn is to the boundary between life and death to ask what actually happens. Which brings us back to the opening vignette and Patti Reynolds, lying on a gurney, deeply chilled, with no detectable brain function. Her story is part of a body of mostly anecdotal evidence on near-death experiences. Eight million Americans claimed to have had such an experience in a 1982 Gallup poll.

A century or more of spiritualism and séances have cast a pall on the scientific study of what happens after death. Scientists were duped and debunkers like Harry Houdini were in a small minority. Elisabeth Kübler-Ross was a noted psychiatrist who claimed that death did not exist. She popularized the association of out-of-body experiences with the point of death, but her credibility was shattered by the revelation that she hired some guy in a turban to have sex with grieving widows under the guise of channeling their dead husbands.

One Dutch study found that 12 percent out of 344 resuscitated patients who had experienced cessation of the heart and/or suppressed breathing function reported a near-death experience, characterized by one or more of the following: a sense of peace, tunnel of light, "life review," conversations with dead friends or relatives, and displacement from the body.[13] These common elements occur without regard to social class, age, race, or marital status.

On the other hand, the accounts vary from culture to culture and it's often hard to prove that they are not false memories. Sensitive MRI methods have shown that there's low-level brain activity in patients who are in a vegetative state. Any person near death is subject to neural noise, cerebral anoxia, endorphin surges, and other extreme neurochemical responses. Dr. Karl Jansen at the Maudsley Hospital in London can reproduce all the main features of a near-death experience with ketamine, a fast-acting, hallucinogenic, dissociate anesthetic.[14]

The "gold standard" in this field would be evidence that a person had seen or known something he or she could not have from the perspective of an immobilized body. Two university physicians, Jan Holden at North Texas and Bruce Greyson at Virginia, hang laptops from the ceilings of their operating theaters such that their screens cannot be seen by the patients. Images play on a random cycle. Patients who claim to have floated above their bodies can report what they saw; to date neither doctor has heard anything intriguing. A three-year study is underway in 25 U.K. and U.S. hospitals to do a similar experiment with as many as 1500 cardiac arrest survivors. It could provide the first convincing evidence that information can pass through the "closed door."

Chapter 2

ALL GOOD THINGS MUST PASS

Benjamin Gompertz is sitting in the pub feeling awkward and out of place. He sips a pint of ale and nervously fingers his belt buckle. Several yards away a group of sailors is deep in a loud and raucous conversation. At a nearby table, a heavyset man is slumped with his head on the table. His tunic marks him out as a bricklayer and his thick fingers are tightly laced around his tankard, even as he sleeps. Gompertz is only 18, but there are even younger boys around him in various stages of drunkenness. The pub is in the heart of Spitalfields, tucked behind the London docklands and within earshot of the bells of Westminster Abbey. The year is 1799.

Gompertz is waiting for someone he first met just one day earlier, a secondhand bookseller who shares his keen interest in mathematics. The man has told him of the Spitalfields Mathematical Society, a group that meets in pubs scattered around the working-class center of the city. The members steadily trickle in, and soon a dozen young men are earnestly huddled around the rough-hewn table, discussing mathematics. There's only one rule, they explain. Anyone receiving tuition from the society has to agree to be a tutor as well. If a fellow member asks a question, his colleagues have to try and find an answer, or pay a penny. This system of peppercorn fines creates an amazing cooperative spirit and a ready exchange of ideas. Gompertz is a Jew so he isn't allowed to attend university. But here he's in his element and he thrives.

By the time he is 40, Gompertz will overcome his humble origins and be elected to the Royal Society. The following year, he will present a paper that applies differential calculus to the estimation of life expectancy. He will work out, for every age of a person, how likely he or she is to die. It will be the first time the seeming capriciousness of death is rationalized and so rendered predictable. The trillion-dollar insurance industry will use Gompertz's law of mortality thereafter.

The Reaper's Scythe

Fending Off Death

If the body is a fortress, it has a busy thoroughfare running through it; our cells are replaced every seven years. As we age, fresh recruits are harder to find so the fortress shows signs of wear and tear. Eventually our moving parts begin to break down and the cell walls are breached by infidels who wreck the genetic machinery. Early humans didn't have the luxury of worrying about the slow grind of aging; their ends came from opportunistic infection or childbirth or the flash of tooth and claw. A short life focuses the mind on living.

Life expectancy at birth has more than trebled in human history.

Our very ancient ancestors only left a few scattered bones to study but it's likely that the first big boost to longevity came from harnessing fire around 500,000 years ago. Cooking meat and fish protected us from many parasites that could kill or shorten life spans. From Neanderthal times through the Bronze and Iron ages that ended in 0 BC there's evidence from pits where the dead were buried. Analysis of skeletons reveals mean ages of 30 and maximum ages of about 45. Throughout the Middle Ages, graves of Anglo-Saxon field workers rarely contained the remains of anyone who survived past 45.

Continuing through the Renaissance, life expectancy at birth was just 25 to 30 years, with the average brought down by high infant-mortality rates. As in every age, it paid to be wealthy. Rulers and members of the English royal court give an unbroken view at longevity from 1000 to 1800. Their life expectancy was 50 years over eight centuries. If they could survive childhood and the rigors of war, their mean life expectancy after the age of 21 was 65, so many wealthy people lived into their 70s and 80s, and a fair number of poorer people too. But let's not feel too superior; a bowhead whale captured in 2001 had a harpoon tip embedded in it that was carbon-dated to 1790. It lived well over 200 years.

There's a fascinating insight contained in the data on longevity. Almost all the gains have occurred in the past 200 years, but they've resulted mostly from a reduction in childhood mortality. In other words, it's a myth that our ancient ancestors endured lives that English philosopher Thomas Hobbes had labeled "nasty, brutish, and short." Medicine was primitive in prehistory, but oral traditions provided the framework for codifying herbal remedies. Awareness and use varied according to the local flora but even a partial list is impressive: Fungi to treat intestinal parasites and as a laxative, rosemary to regulate blood pressure, curare as a muscle relaxant, foxglove to reduce swelling and as a stimulant, and motherwort to treat asthma. Clay was used to form simple splints to set bones.

Of course some of the remedies make us cringe. Certain tribes in the Americas used large ants for an early form of stitches. The ant was held over the wound and bit into it, then the body was broken off so the pincers would remain holding the skin tight. Cultures across the world practiced trepanning or trephination. In this stomach-churning procedure, flint tools were used to cut a five-centimeter diameter hole in the skull. Trepanning was apparently used to treat severe headaches and epilepsy, or simply to let evil spirits out. Amazingly, regrowth of the hole and the fact that people kept the skull disk with them as a lucky charm shows that most people survived this procedure, which was of course performed without anesthetic.

Civilization was not the panacea for human health that most people imagine. Ethnographic descriptions of modern hunter-gatherers and the archeological record of extinct tribes show that they had higher caloric intakes and more balanced diets than modern citizens of third-world countries or the urban poor in industrialized countries. In turn, the main features of the hunter-gatherer diet—high bulk, low fat, high potassium, low sodium, and low calories relative to other nutrients—acted to protect them from the rate of diabetes, circulatory problems, degenerative disease, and strokes that afflicted sedentary populations. Nomads tended to have lower "loads" of parasites than city dwellers and transmitted diseases such as tuberculosis, malaria, and bubonic plague had less purchase among small, isolated populations. Finally, fixed civilizations were just as afflicted by catastrophic resource failures as migratory humans until fairly recently. Being civilized is overrated.

Medicine to the Rescue

The Hippocratic Oath begins "I swear by Apollo Physician and Asclepius and Hygeia and Panacea and all the gods and goddesses, with them as my witnesses, that according to my ability and judgment, I will fulfill the following oath." Panacea was the goddess worshipped by the sick who hoped to heal and Hygeia was the goddess worshipped by healthy people who wanted to stay that way. As science has supplanted 2400-year-old superstition, the noble goal of the oath has been realized.[1]

The gains in life expectancy in the age of modern medicine have been dramatic. After centuries when mean longevity fluctuated between 30 and 40 (with a major dip during the Black Death), it started

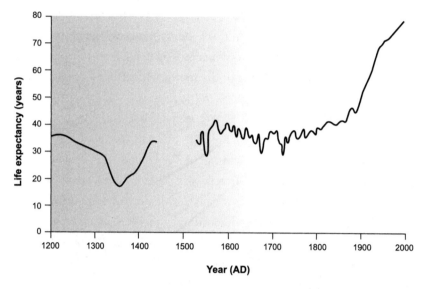

Figure 2.1. Life expectancy at birth of male landowners in England between 1200 and 1450 AD, and for the both sexes in England and Wales between 1541 and 1998. The average of 35 for most of this time was no higher than during Roman times. The doubling of life expectancy in the past century is mostly due to reductions in infant mortality and, recently, to success in combating some of the diseases of old age.

to steadily rise in the Western world during the mid-nineteenth century (Figure 2.1). Germ theory held that transmitted microorganisms were the cause of many diseases. Louis Pasteur refuted the alternative idea of spontaneous generation, and German physician Nobel Prize–winner Robert Koch provided convincing proof for the germ theory, but it took decades for these ideas to translate into medical practice.

Luckily, even before these insights, a Hungarian obstetrician named Ignaz Semmelwies guessed that the fever that killed one in three women during hospital childbirth was transmitted by the doctors, who often went from an autopsy to deliver a child without washing their hands. Better hospital practice reduced these unnecessary maternal deaths, though only to the level of 1 percent that had always been the norm when babies were delivered by experienced midwives.

Steady improvements in standard of living, diet, and medical practice pushed life expectancy up more steeply throughout the twentieth century.[2] In 1900, 10 percent of babies didn't make it past the age of one and 20 percent didn't make it past the age of five. By 2000, those percentages were 0.7 percent and 0.8 percent, respectively. Improved medical care was good news for moms too. Maternal mortality dropped from 1 percent of births leading to the mother's death to 0.01 percent or only 1 in 10,000.

People born in 1900 could expect to live 50 years. A child born in 2000 can expect to live 80 years. There are two threads to this story. One is the massive success in fighting infectious and communicable diseases. Tuberculosis killed 200 people per 100,000 population in 1900 but was essentially eradicated as a cause of death by midcentury. Pneumonia and flu together killed similar numbers in 1900 and were reduced by a factor of 10 by the century's end. In the United States, as in the rest of the industrialized world, cholera and smallpox are now unheard of. Driving might feel hazardous, but it's actually safer than it used to be. In 1930 the death rate from road accidents was 25 per 100,000; now it's 40 percent less, despite 10 times more cars on the road.

The second thread is the transition to different modes of disease. In the mid-twentieth century, falling death rates for communicable diseases crossed rising death rates for chronic diseases. The major causes of death among Americans are now heart disease and cancer, with rates of about 300 and 200 per 100,000 population, respectively.[3] The old killers were invisible, but at least there was a single causative agent. The new killers have a complex web of causes and risk factors.

We've become victims of our success in combating the relatively easy diseases. The rate of increase in longevity has slowed. Life expectancy at birth increased by 18 years from 1900 to 1950 but only by 8 years from 1950 to 2000. U.S. death rates from malignant cancer rose from 65 per 100,000 in 1900 to 200 per 100,000 in 2000, and death rates from cardiovascular disease stayed steady at about 340 per 100,000 population (Figure 2.2).

However, crude mortality figures conceal a hard-fought success. Death rates for chronic diseases are 1000 times higher at age 80 than they are at age 20, so as life spans are extended, more people will succumb to these diseases. In the U.S., the age-adjusted mortality rates for heart disease have been falling since the 1960s, spurred by conscious-

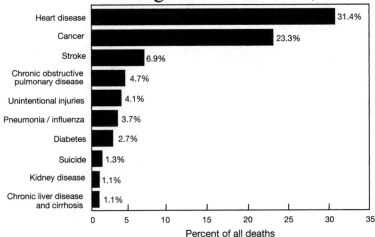

Leading Causes of Death, 1997

Cause	Percent
Heart disease	31.4%
Cancer	23.3%
Stroke	6.9%
Chronic obstructive pulmonary disease	4.7%
Unintentional injuries	4.1%
Pneumonia / influenza	3.7%
Diabetes	2.7%
Suicide	1.3%
Kidney disease	1.1%
Chronic liver disease and cirrhosis	1.1%

Percent of all deaths

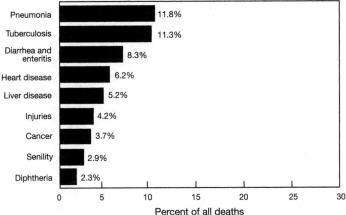

Leading Causes of Death, 1900

Cause	Percent
Pneumonia	11.8%
Tuberculosis	11.3%
Diarrhea and enteritis	8.3%
Heart disease	6.2%
Liver disease	5.2%
Injuries	4.2%
Cancer	3.7%
Senility	2.9%
Diphtheria	2.3%

Percent of all deaths

Figure 2.2. Dramatic changes in the leading causes of death in the United States between 1900 and 1997 were the result of the defeat of traditional infectious and communicable diseases, and the rise of more insidious killers like heart disease and cancer, which have a complex web of causes. Part of the shift can be simply ascribed to increased longevity. The death rate from accidents is stable; at least we aren't getting any clumsier.

ness of the dangers of smoking and the benefits of exercise and proper diet. Age-adjusted mortality rates for malignant cancers peaked in the mid-1980s and they've been slowly declining since. The trillions of dollars spent on treatment and prevention are actually having an effect.

There's a wide range in life expectancy worldwide which is caused by inequalities in wealth and access to medical care and good nutrition.[4] Poverty and the ravages of AIDS keep the life expectancy in Botswana and Swaziland around 33 to 34 years. At the other extreme, the best places to live are microstates Andorra and San Marino, where people live 82 to 83 years. In Russia, men rank 166th in the league table with a longevity of 59 years, which is 14 years less than Russian women. Sometimes the variation is dramatic within a country; life expectancy for men in the Glasgow slum of Calton is 54 years, 28 years less than in affluent Lenzie, only five miles away. Something unspeakable must have ended up in their haggis.

Taking No Chances

The odds that you and I will die are, hmm, let's see . . . carry 5, take the square root, convert to base 2 . . . nearly there . . . OK, got it: 100 percent. But the most likely manner of our passing is something we can influence, so it's worth looking at the most and least probable outcomes.

Worldwide, heart disease kills about half of all people, infectious and respiratory diseases kill about 30 percent, cancer kills 12 percent, injuries kill 6 percent, and everything else is under 5 percent. But these statistics conceal the fact that there are two "worlds." In industrialized countries, cancer and heart disease are dominant causes of death. In the developing world, AIDS is the biggest killer, followed by lower respiratory infections and then heart disease. Incredibly, diarrhea kills more people in the third world, mostly children, than cancer kills in the West. Almost all of the major causes of death in poor countries have been reined in or even eliminated in rich countries.

The national variations are fascinating and often informative. Heart disease kills proportionally two to three times fewer people in southern Europe than northern Europe, which can be attributed to the Mediterranean diet. Cancer rates are high in Denmark but low in nearby Finland. If you want to be surrounded by healthy livers, go to the Netherlands, Norway, or New Zealand. Driving is most dangerous in Portugal and Russia, safest in Sweden and England. Steer clear of Russians and Hungarians, who kill themselves seven times more often than san-guine Spanish and Portuguese.[5]

Granted, fear is irrational, but we seem to have inverted all its logic. Driving is by far the most dangerous thing we do daily yet few people maintain a suitable sense of the threat. Driving is the leading cause of death among U.S. teenagers. On the flip side, many people are afraid of flying, yet the odds of dying are 1 in 19,000, so you would have to fly every week of your life before it approached the risk of your dying of cancer. Fear of a terrorist attack is just silly.[6] For every American killed by a terrorist, 10,000 are shot to death by nonterrorists, and most of those nonterrorists are people they know, usually relatives.

Some cold, hard facts: Cold is more dangerous than heat. Tooth-picks are more dangerous than lightning. Pedestrians are more dan-gerous than fire. Beds are more dangerous than ladders. Snakes are more dangerous than terrorists. Flammable nightwear is more dan-gerous than a bee sting. Coconuts are more dangerous than sharks.[7]

Here's some practical advice based on this welter of statistics. The American Medical Association has shown that most deaths in the U.S. result from modifiable behaviors. If everyone ate well, exercised, and avoided smoking, the mortality rate would drop by a third.

That's simple, but a little dull, so let's drill down the list a bit fur-ther and see what other lifestyle changes you might consider. Acci-dental injury leads to death with 1 in 36 odds, and falling down is also quite dangerous, so once you've done your exercise stay quietly inside and sit in a chair. The next most dangerous thing is you—entertain happy thoughts to fend off suicide. A set of natural dan-gers can be avoided by choosing the right place to live. A nice place like Muncie, Indiana, will immunize you from tornados, hurricanes, earthquakes, tsunamis, and drowning. Housing is cheap there too. Your odds of dying by legal execution are 1 in 60,000, but much lower if you behave yourself.

Going further risks tilting into obsession, but what the heck. Water's hazard is inversely proportional to the amount. So bathtubs are more dangerous than swimming pools, which are more dangerous than the ocean. Be very careful of that cup of tea or coffee. Wear an armored, asbestos, nonconducting bodysuit to handle additional situations like dog attacks, venomous bites, electrocution, and immolation. Gambling can be a problem too. National Public Radio reported that New Yorkers were 17 times more likely to die crossing a bridge into New Jersey to buy a ticket for that state's major lottery than they were to win that lottery. It's another reason to just sit in a chair.

Finally, and this isn't advice since you have no control over it, hope you're a woman. Women live longer than men. Their major threat—death during childbirth—has been ameliorated by better medical attention, and they mostly avoid dying in wars. Men are their own worst enemies. We drink more, smoke more, take more drugs, die more often behind the wheel, and kill each other and ourselves more often. As a result, the life expectancy gap between men and women has been growing.

Beating the Odds

Pausing Death

People struggled for centuries with how to deal with the uncertainties of life, including the timing of death. Charity has always existed as a safety net, but it's imperfect and carries a social stigma. The ancient Greeks invented pensions and the Romans formed societies to pay for burial expenses. By the Middle Ages, you could pay a lump sum to a monastery and live there for the rest of your life, if you could handle solitude and incessant praying.

The science of dying began in 1662 with John Gaunt, a London draper who discovered regularities in the patterns of death among groups of people despite the very large uncertainty in the future lifetime of any single person. Which brings us to Benjamin Gompertz, whom we met in the opening vignette. Gompertz teased mathematical regularity from the capriciousness of death. He built on the work of fellow Londoners: comet-hunter Edmund Halley and French refugee

Figure 2.3. Benjamin Gompertz (1779–1865) was the world's first actuary. His law of mortality, used by insurance companies worldwide, is a refinement of a demographic model of Thomas Malthus's. He was self-taught because as a Jew he was denied admission to university.

Abraham de Moivre, who as an old man noticed he was sleeping slightly longer each day, and poignantly and accurately predicted the date of his own death. Gompertz found that what he called the "force of mortality" doubled every seven years after the age of 20. This precise formulation made him the world's first actuary (Figure 2.3).

Thinking about the fact or manner of your death is bad for the health—there's a neat paradox. Tolstoy spent his last years in the fevered grip of existential angst, and it clearly hastened the novelist's demise. Yet people who calculate mortality do well. *The Wall Street Journal* noted several years ago that actuaries have among the very best jobs, with low stress, high pay, and a cloistered work environment.[8] At the other end of the spectrum are lumberjacks and fishermen; if you can choose, always work with numbers and not nature. Gompertz comfortably exceeded statistical expectations, living to the ripe age of 86.

Actuaries don't dwell on death; their skill is to calculate risk in many situations. They're used to cruel jokes—an actuary is someone who wanted to be an accountant but didn't have enough personality. The 2002 movie *About Schmidt* cast Jack Nicholson as a cantankerous actuary who had just lost his wife. The Society of Actuaries was upset enough to post this official response to the movie on their Web site: "The portrayal of actuaries as math-obsessed, socially disconnected individuals with shockingly bad comb-overs is 97.28892% incorrect." We astronomers may languish behind actuaries in an alphabetical list of professions but we outshine them on the silver screen, with Daryl Hannah (*Roxanne*, 1987) and Jodie Foster (*Contact*, 1997) as recent exemplars.

The Inexorability of Aging

Getting old sucks. That obdurate fact is compounded by the suspicion that youth is wasted on the young, who take it for granted. At least in middle age, the rate of decline is gentle, such that I'm able to ignore the slow loss of body plasticity and a gradual erosion of acuity of my senses (or maybe I can't remember . . .). Gains in longevity in the past century have been impressive, but there are diminishing returns as people are increasingly picked off by degenerative illness and cancer. As Yeats said, "Things fall apart; the center cannot hold." We've been sucker punched the way Tithonus was by Zeus when he was granted immortality but not eternal youth; as a result he became debilitated and demented. Aging is like a rocky road that ends in a wall; we can move the wall back a bit but we can't knock it down (unless you're a die-hard dualist).

Aging is a mysterious process. Nobody has yet found a fundamental biological process that controls it and the complexity of cell chemistry and population genetics make identifying the underlying causes very difficult. Perhaps the biggest problem is separating cause and effect. For example, there's no direct evidence to back up the very tempting assumption that disease causes old age and death.

Theories of aging fall into two categories: either it's a feature or a bug. The bug or "defect" school holds that aging is the result of unalterable deterioration at the cellular level. The body is a mechanical system like a car, where entropy or oxidation or other types of molecular damage cause things to fall apart. We can treat individual manifestations like arthritis, cataracts, and heart disease—analogous to fixing the engine or the brakes or getting a new paint job—but the cause is untreatable. To people holding this view antiaging research is misguided because the degradation is inevitable.

Free-radical theory says that aging is caused by atoms and molecules with unpaired electrons. Oxidizing agents are created by the cell's own metabolic reactions and they're present in the environment, but they are also unstable and very reactive so they inflict substantial chemical damage within cells. Aging is supposed to be the cumulative effect of molecular damage on the machinery of the cell; in effect, our bodies "rust" from within. A lot of people must be convinced; the market for antioxidant supplements like vitamins C and E was $3 billion in 2007. But while free radicals play a role in particular age-related pathologies like cataracts, controlled experiments with antioxidants on mice have yielded ambiguous results and there's no evidence that free radicals cause aging in general.

The alternative idea is that aging is a built-in feature of evolution, a genetically regulated or programmed process. Identifying the genes responsible for aging has proved difficult. However, the landmarks of development, reproduction, and aging occur in the fixed proportion to the entire life span in many mammals, which suggests that aging is a result of the genetic program that controls development. There are rare disorders where a single rogue gene causes accelerated aging in humans, in one case leading to severe heart disease in teenagers. In other cases, a single gene can be disabled in worms and mice to give an increase in longevity by as much as a factor of 6 (Figure 2.4).

Aging seems to go against Darwin's theory of evolution. Why would evolution favor a process that increases mortality and decreases the reproductive capacity? One possibility is that genes that have bad consequences later in life can be positively selected as long as they have benefits early in life. Another is called the "disposable soma," where in a world full of external dangers it's better to devote more resources to reproduction than repair. While it's important to explain the general rule that organisms get less fit as they age, we can learn much from organisms that thumb their noses at the rule.

Almost Immortal

Aging isn't inevitable or universal. As an example of the subtlety of aging, take the Hayflick limit. Named after a genetics researcher, this was the discovery that a normal cell in culture divides about 50 times before dying. A limit arises because chromosomes are all capped by

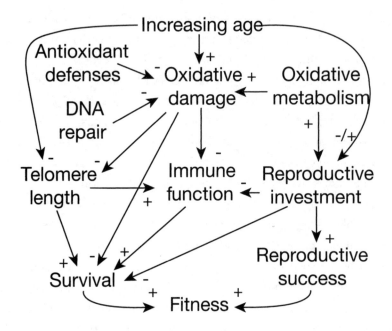

Figure 2.4. The complex interplay of factors involved in aging is shown in this diagram that illustrates both the positive and negative effects on fitness, which manifests as longer survival and reproductive success. It's difficult to control or even understand the factors in one species, but when more than one species is involved, variations of environment come into play.

a repetitive chunk of DNA called a telomere. Telomeres are consumable buffers of information and when they're used up the cell can no longer reproduce, seeming to support the idea that death at the cell level is preprogrammed. But they're also useful because they stop chromosomes from being fused and rearranged rather than just duplicated, which is major cause of cancer.[9]

The phenotype of human aging is one where any system, tissue, or organ can fail. When the "weakest link" fails, death is the result. This is true for centenarians as well; nobody simply dies from "old age." As we age, we become more susceptible to certain diseases, but aging is not just a collection of diseases. The focus on specific pathologies can also distract researchers from the common mechanisms of aging that encompass all of them.

Some species do not seem to age. Leach's petrel, a small sea bird of the northern oceans, has telomeres that lengthen with age so its cells are essentially immortal. The liver, lungs and kidneys of a centenarian turtle are indistinguishable from the organs of its teenage counterpart. Turtles have the ability to control the pace of their metabolism. If they didn't succumb to disease or predation they might not age. Rockfish, sturgeon, and lobsters also defy time's passage; the older individuals aren't weaker, less agile, more susceptible to disease or less fit in any way than the younger specimens. All of these species continue to grow and become more fertile as they get older. Longevity within one genus of rockfish ranges from a dozen to 200 years, although they all inhabit the same environment.

For true immortality we turn to a transparent jellyfish about six millimeters across, with a bright red stomach and 80 to 90 tentacles. *Turritopsis nutricula* are found throughout the world's oceans. Unique among members of the animal kingdom, they reverse their life cycle after becoming sexually mature, returning to the polyp state. They can repeat the cycle infinitely, cheating death.

How do these animals do it, and what is the selective advantage in long life? Lobsters use a special enzyme that prevents the decay of telomeres, while somehow avoiding cancer. Rockfish have created their own antioxidants. The reason is often unknown. But the natural selection mechanisms for extreme longevity are understood. Arctic clams and bristlecone pines each live in crowded but stable conditions where new opportunities to mature are rare. It pays the individual to outlive its neighbors, so its seedlings or larvae have a place to grow. This sets up an antiaging arms race.

The debate over aging as a bug or a feature of biology has no simple resolution. If aging is the result of accumulated damage at a molecular level, then why does a mouse-sized lump of flesh wear out hundreds of times faster than a human-sized lump of flesh? The answer could only lie in better repair mechanisms in humans but those haven't been found. On the other hand, a gene can make its carrier stronger, smarter, and more resistant to disease, but it can't eliminate all the external causes of damage and decay in the environment. But the existence of ageless animals and identifiable genes that regulate aging makes researchers cautiously optimistic that we might one day make time stand still and extend human life spans far beyond present levels.

Dust to Dust

Death and Disorder

Beginnings and endings are part of the texture of life and the rhythm of the universe, but time itself is deeply mysterious. Our unmistakable sense of time's arrow emerges from basic physics. Individual atoms have no clocks and particle reactions can proceed forward and backward in time, yet ensembles of atoms manifest forward-flowing time as an emergent property. We can imagine alternate universes, hypothetical yet physically plausible, that don't have the flow of time; our universe happens to have a story to tell.

But the story would be sterile if there were no ears to hear it. Life is one of the most remarkable attributes of the universe. Space is mostly an almost-perfect, absolutely frigid vacuum. Biology is only possible, as far as we know, on or near the surface of a planet or a moon at a particular distance from a suitable star. Lives are fleeting—ours are to the age of the universe as a hummingbird wing beat is to a fortnight.

The evolution of the universe can be described in terms of an overall progression to higher entropy. Terms associated with entropy are disorder, chaos, randomness, and loss of information. On the face of it, biology seems to violate this progression; when something goes from nonliving to living it becomes more ordered, has more structure, and encodes more information. But the cell or the organism is not the whole system. To talk properly about entropy and biology, the flow of energy in and out of life has to be considered.

Living creatures are able to keep their entropy low because they take in energy in the form of food. Their order is gained at the expense of disordering the nutrients they consume. In photosynthesis, the order of biomolecules is created from the chaotic energy of sunlight. Rather than thinking of entropy as disorder, it's more useful to think of it as heat, which is the most chaotic form of energy.[10] In photosynthesis, sunlight drives a cyclic chemical reaction and heat is produced along with glucose. We and other animals radiate heat into the environment, payback in increased entropy for the energy we use to live and grow.

Earth swarms with life and even though the Sun can seem irrelevant to a city dweller, it's the base of a pyramid of life. The biosphere is powered by sunlight so life on Earth is an intermediate temperature way station between the concentrated hot radiation from the surface of the Sun and the cold, diffuse bath of low-energy photons in intergalactic space (Figure 2.5).

Entropy provides us with a microscopic framework for understanding the passage of time. There's a random loss in the fidelity of molecular interactions, which accumulates to steadily overwhelm the repair and maintenance systems. Alzheimer's disease is an unfortunate example. Increasing entropy manifests as a network of interacting age-related changes in the brain, which drives neural and cognitive decline in the elderly. A simple model of entropy generation in our metabolism can closely predict male and female life spans.

Entropy is implacable, but if it's a jailer, we find ourselves pris-

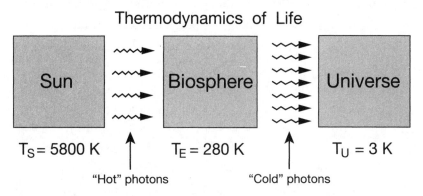

Thermodynamics of Life

$T_S = 5800 \text{ K}$ $T_E = 280 \text{ K}$ $T_U = 3 \text{ K}$

"Hot" photons   "Cold" photons

Figure 2.5. The continued existence of the biosphere appears to violate universal tendency toward increasing disorder. The source of negative entropy for life on Earth is the organized and directed energy of light from the Sun. The photons are degraded in the biosphere at an intermediate temperature and released into space as thermal radiation of a lower temperature, or heat. So "source" photons at 5800 Kelvin are turned into "sink" photons at 3 Kelvin. Also, sunlight is directed and only a billionth of it is intercepted by the biosphere but the planet radiates heat in all directions, increasing the entropy.

oners with comfy living quarters and a lot of latitude. The environment has abundant free energy for creating new structures, and organisms can forestall decay and death for a long time with good adaptation and repair mechanisms. Just try to ignore the sagging body parts.

Frozen in Time

As I write this I am 53 years old. I keep fit and eat reasonably well and I'm far too embedded in the texture of life to think about how it might end. Although I have a vivid imagination, it's not nearly good enough to encompass the cipher of my nonexistence. People I know talk about heaven or cycles of birth and death or continuation as an intangible, ethereal entity. I listen with polite interest; as an agnostic I've no basis for either believing or disbelieving these ideas.

If I'm lucky I'll thumb my nose at entropy for another 30 or more years. But each year about 60 million people die and it seems like a waste. What if they'd been able to continue to be with their families, appreciating music and the stillness of dawn, continuing to work and be productive? On the other hand, if all we are doing with medical technology is adding low-quality declining years to the end of the life span, perhaps we should take the energy we put into staving off the inevitable and invest it in vividly living the time when we have all our faculties. Death is bad, but prolonging the inevitable may be worse.

To fans of cryonics, that's quitter talk. Until the 1970s you had two choices when you died: rot in the ground or go up in smoke. Now you have a third: be frozen and hope people with future technology thaw you out. For a cool $150,000 you can be frozen to the temperature of liquid nitrogen, $-196°C$ ($-321°F$), and half of that cost preserves just your head. The latter creates a ghoulish image, but it makes sense because cryonics is based on the premise that memory and identity are stored in the tissues of the brain; in the future they'll surely be able to graft on a really nice new body.

Cryonics has received attention out of all proportion to the number of people affected. Less than 200 people—sometimes minus their

heads and sometimes plus their pets—are on ice at two hi-tech facili-
ties, one run by Alcor in Scottsdale, Arizona, and the other run by the
Cryonics Institute in Clinton Township, Michigan. Each company has
a "waiting" list of about a thousand. Cryonics suffered a setback in 1979
when a lack of funds caused the "meltdown" of nine bodies. Oops. It
seems that patients who sign on for the deep freeze are placing inordinate faith in the continuity of utility companies. Both cryonics companies claim to have solved the problem of cell damage caused by ice
crystals formed during the cooling.

So far the undead are housed in drab suburban warehouses. Famed
architect Stephen Valentine, designer of Washington's Holocaust
Museum and the new Long Island Railroad Station in New York, plans
to change that. His Timeship will rise like a modernist cathedral out of
six landscaped acres and hold 50,000 frozen optimists. A fine mist will
drift across its central plaza and angled mirrors at the perimeter will
reflect the sky above. He says, "It's a Fort Knox of biological materials,
and it could be Noah's Ark to the future."

Who are these quixotic emissaries? Most prefer to remain anony-
mous but those who don't are mostly men, a mixture of entrepreneurs
and techies. A popular urban legend holds that Walt Disney was frozen
in 1966 and is interred under the Pirates of the Caribbean ride. In fact,
he was cremated and his remains are in Forest Lawn cemetery. Dick
Jones, a writer for *The Carol Burnett Show*, is in the Alcor facility, and
Alcor is holding on to his Emmy until he can retrieve it. The most celebrated subject of cryonics is the last man to hit .400 in the majors. Ted
Williams—the "Splendid Splinter"—is stored upside down in Scottsdale. A former vice president of Alcor suggested, not entirely in jest,
four reasons why people come to them: They're so appalled by death
they'll try anything, they're extreme narcissists, they have no answer to
death, and they enjoy being part of a ridiculed minority.

The adult brain is an electrochemical network of 100 billion neurons and 60 trillion synapses. It's a leap in the dark to imagine that
it can be reanimated by any future technology. Personal survival is in
doubt but not the survival of humanity because sperm and eggs can be
frozen without adverse effects.[11] Successful pregnancies have resulted

from embryos frozen for nine years and children born from those embryos, called "frosties," show no increase in birth defects or developmental abnormalities. Fans of cryonics admit it's a leap in the dark, but say there's little downside except the cost; if you're already dead, what's the worst that can happen?

Not Dead Yet

The premise of cryonics may be misguided, but the practice is part of renewed debate on the definition of death. Last chapter, we looked at the idea of survival beyond the body, but the point at which the body gives up is also of interest. The son of the founder of the movement, David Ettinger, has said, "Death is just the point at current technology when the doctor gives up. It's a legal definition, not a medical one." Since 1968, most developed countries have adopted a definition of death based on extinction of activity in the brain, rather than the heart or any other organ.[12]

The heart is the engine that keeps the brain alive with oxygen. When the heart stops, the brain is considered to be dead after five minutes with no electrical chatter. However, the increased demand for organ transplants is causing doctors to alter their procedures to get organs while they are still "fresh." Often, they wait for a dying patient's heart to stop beating, but instead of waiting the few minutes for the brain to die as well, they anticipate the inevitable and declare the patient dead immediately so they can harvest organs. Death is based on a decision not to resuscitate rather than the *impossibility* of resuscitation.

If that makes you uncomfortable, there's worse. Standard medical practice following cardiac arrest may actually accelerate the death of the patients doctors are trying to save. About 250,000 times a year, someone's heart stops beating in the United States. Survival is mostly luck, depending on the response time to a 911 call and the availability of local CPR. If it happens to you, your likelihood of dying is 95 percent. As we saw in the last chapter, some researchers are questioning whether death is 100 percent fatal, but let's assume for the sake of argument that it's something to avoid at all costs.

If your heart stopped, within 20 seconds the hundred billion neurons in your brain would use up their residual oxygen and all electrical activity would fade away. If you got to a hospital, doctors would try to revive your heart with a defibrillator and if that worked (and even if it didn't) they'd put you on oxygen. However, recent research shows that cell death is a long and complex process. Both neurons and heart cells can survive without a supply of blood for hours, perhaps as much as a day. But after more than five minutes without oxygen they die when the oxygen supply is *resumed*. Apparently, the sudden infusion of oxygen to cells in a shut-down state triggers the self-destruct mechanism that usually protects against cancer.

Here's what researchers think is happening. The voracious demands of the heart and brain for oxygen mean that oxygen starvation is indeed fatal. But dying at a cellular level is complicated, and it takes hours, perhaps days, to be complete. Researchers like Lance Becker, at the University of Pennsylvania's Center for Resuscitation Science, think answers lie in mitochondria, the tubular structures within cells where oxygen and glucose combine to give our bodies the energy they need to function. The process of cell death begins with mitochondria, which control apoptosis, the self-destruct mechanism for cells that aren't needed or are damaged. Cancer cells proliferate by shutting off the mitochondria so the self-destruct mechanism is disabled, and cancer researchers are looking for ways to switch the mitochondria back on. Doctors like Lance Becker are trying to do the opposite, preventing cells that have been damaged by lack of oxygen—but not beyond repair—from committing suicide.

A better procedure in the case of cardiac arrest seems to be late and gradual application of oxygen but immediate and aggressive cooling of the patient. It's long been known that people who fall through ice can survive unexpectedly long immersions in water. Napoleon's surgeon general noticed that wounded infantrymen left on the snowy ground had higher survival rates than those that huddled near a fire. Nobody knows how it works, but slowing the metabolism by hypothermia and avoiding cellular suicide by excessive oxygen buys enough time to let many heart attack victims recover.[13]

Hope is not a strategy. But thwarting the great majority of deaths by heart attack, and giving people an extra decade or so for the kinks to be worked out in cryonics, which bridges us to the longer term future when we can be taken off ice and slapped on the fanny to get us jump started again, much as we were when we were brought into the world, now that's a plan we can live with.

A Loan Repaid

The big view of our birth and death is that they are small transactions in the global economy of the biosphere. Nature is parsimonious. Most of an organism's ingredients were once part of a variety of long-dead organisms and they will eventually get recycled, perhaps into species that don't yet exist. Think of your atoms as hand-me-downs.

As we've seen, you're not as old as your birth certificate. A few years ago, scientists figured out how to tag DNA with carbon-14. Molecules flow in and out of a cell but DNA doesn't, therefore radioactive carbon enrichment can be used to figure out a cell's age. The main ingredient of life is water. All cells contain water, although the proportion ranges from 5 percent in dormant seeds to 95 percent in jellyfish and young plants. About 60 percent of human weight is water, and the typical water molecule cycles through your body in a few days. Note, however, that reducing water intake isn't a sensible weight-loss strategy; a dehydrated body actually craves food. Without replenishment—any healthy adult needs to drink 2.5 liters a day—death results in 10 days or less.

The other essential ingredient for life is carbon, which makes up 18 percent of human weight. Apart from water, the major components of human nutrition—fats, carbohydrates, proteins, and vitamins—all contain lots of carbon. Carbon atoms are eternal, but what's their passage through the narrative of life?

There are three very different timescales for the cycling of carbon. The longest is the hundred million years or so that a carbon atom might

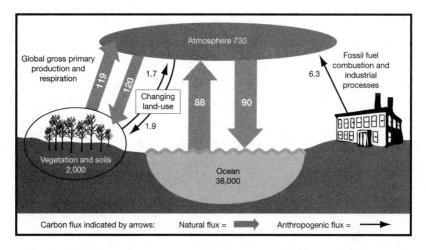

Figure 2.6. The carbon cycle, shown graphically as fluxes of carbon, in units of billions of metric tons. Stockpiles contained in the ocean, land, atmosphere, and air are large, and the most rapid exchanges occur in the biosphere. As we have learned recently to our cost, small changes in the exchange rate of carbon into the atmosphere can have large impacts on global climate.

be trapped in the crust before tectonic activity or volcanism liberates it to the atmosphere. Two thousand times less carbon is interred for as long as 100,000 years in the deep oceans. An even tinier amount of carbon, 0.001 percent of the stockpile in the crust, is part of a three-way exchange between the atmosphere, plants and animals on land, and microbes in the surface layers of the oceans. This transactional fizz is responsible for the ephemeral layer of the biosphere, with a carbon atom changing its situation every 1 to 100 years (Figure 2.6).

There are 80 billion metric tons of carbon in the sum of all humans on the planet. That's barely 0.01 percent of the carbon in living organisms and a miniscule fraction of the carbon available in the biosphere. We're dwarfed by such numbers but we can see in them affirmation of the planet's extraordinary ability to form creatures like us.

The second layer of narrative is cosmic. We merely borrow our atoms from the biosphere for a short span and at the end—bury or burn—we will return them. The backbone of life is carbon, and ours was forged in the white heat of the centers of stars and burped into space long ago and far away. The universe is 13.7 billion years old, and some of our carbon might date back to the first generations of massive stars over 13 billion years ago, but the great majority of our carbon atoms are 2 to 8 billion years old. Carbon atoms are made in stars with helium in their cores if the temperature is 100,000,000°C (180,000,000°F) or greater. An extraordinary "oven" is required to cook life's key ingredient.

Many carbon atoms are entombed forever in the cores of massive stars when they die, but some fraction is ejected by the paroxysms such stars undergo late in their lives. Space is filled with a "smog" of carbon-rich atoms, molecules, and small particles, an inventory that includes carbon monoxide (CO), carbon dioxide (CO_2), methane (CH_4), form-aldehyde (H_2CO), cyanogen (CN), carbon monosulfide (CS), carbo-rundum (SiC), acetylene (C_2H_2), methanol (CH_3OH), benzene (C_6H_6), and even larger molecules, ranging up to buckyballs (C_{60}). When a new star forms, this carbon is included in the material that forms dust, dirt, rocks, planets, and, eventually, us.

The story ends there unless you opt for space burial. After crema-tion, a 70-kilogram adult is reduced to about 5 kilograms of carbon-rich ash. For $2500, Celestis, Inc., of Houston will pack a mere gram of your remains in a lipstick-sized tube and launch it into Earth's orbit. However, drag from the remnants of atmosphere at that height will eventually cause the ashes to return to Earth. If you want to truly leave the gravitational influence of the Sun, Celestis plans a service starting in 2011 that will allow you to potentially contribute to new life, starting at $12,500. Up to now, the only person with that special fate is Clyde Tombaugh, who will visit his (recently demoted) planet Pluto on NASA's *New Horizons* in 2015 and then continue into deep space.

The rest of us have to return our borrowed atoms, but we can reflect on their extraordinary journey. We each host 4×10^{27} carbon atoms. Atoms are gregarious, so the number of histories that led them to be part of our bodies may be considerably smaller, but it's still a huge number. We can each try to visualize the myriad pathways in space and time that have led them to become part of a breathing, thinking assem-blage that can puzzle over its own origin and fate.

Chapter 3

THE FUTURE OF HUMANITY

He was told they died laughing. Vincent Zigas considered this unlikely fact as he wrote in his journal by flickering lamplight, keeping a watchful eye on the large insects that crawled in his tent. "Was it an invisible miasma that killed these people? An unknown epidemic influence of atmospheric or cosmic or telluric nature, all pervading, inexorable, sneaking into them, poisoning them, killing them?"

It was 1950, and Zigas was an Australian doctor working in remote parts of Papua New Guinea. The mountainous landscape held over 700 tribes, each with their own language and many of which were unaware of each other's existence. At the dawn of the Space Age there were Stone Age cultures eking out an existence in cloud-draped rainforest. In one village he encountered members of the Fore tribe in various stages of a terrible illness. Some suffered from slurred speech, unsteady gait, and nonstop shivering. Others could no longer walk; they lay in bed and jerked their limbs, with mood swings and bursts of random laughter. A final group was incontinent and covered with ulcers. Progression though the stages of the disease took a year. They called it kuru, the Fore word for tremor. Almost all of the victims were women. As far as Zigas could tell, the disease was always fatal.

Zigas wasn't able to make the connection between the suffering he saw and the funerary rituals he witnessed, where maternal kin of a recently deceased villager would dismember the corpse, stripping away the flesh and removing the brain and internal organs. He didn't associate it with the fact that women would feed children and the elderly morsels of those internal organs, or serve a pale gray broth made from the brains.

The connection was made later, by Carleton Gajdusek, who later won a Nobel Prize for his work. Gadjusek deduced that kuru was not genetic, and was not viral, but was related to disease that had been known for hundreds of years in sheep, called scrapie. The infectious agent itself was not isolated until the early 1980s by Stanley Prusiner, who was also awarded a Nobel Prize for his insight. The brains of humans who die in this way are riddled with holes, like Swiss cheese. The Fore stopped dying from kuru after they stopped practicing cannibalism.

The cause of kuru in humans, scrapie in sheep, and their bovine cousin "mad cow disease" is a prion—a pathogen so small, so seemingly inconsequential, it's shocking it can cause a fatal and untreatable human disease. A prion is a misfolded protein. Biologists were surprised to learn that a protein could be an infectious agent because the "central dogma" of biology holds that nucleic acids like RNA and DNA are required for the transfer *of genetic information. Imagine a single misfolded sock in one drawer that could cause death to all the inhabitants of a house.*

The Fate of Species

A Rose Is a Rose

Is a rose. To be grounded in the idea of species, where better to start than something familiar, something that by any other name would smell as sweet. Roses are endemic to the Northern Hemisphere, with fossils going back 35 million years to the Oligocene epoch. They were gracing the planet long before we evolved to admire them. Wreaths of roses were found in the Egyptian tombs, and they feature in frescoes in Crete from 1700 BC, during the heyday of the Minoan culture, and in Cuneiform tablets from Mesopotamia. The Greeks and Romans used roses in their festivals, both sacred and profane.

Imagine I give you two roses. They differ in color, scent, the size and shape of the petals, and the nature of the leaves and thorns. Are they different species? Deciding would be difficult because botanists tell us there are over 100 species, yet members of some species show more variation than samples that are declared to be separate species. When we got our hands and green thumbs on them, defining species became an even more thorny issue. In the late 1700s, hardy perennial roses that the Chinese had grown for thousands of years were first mated with European roses that only bloom once a season. The subsequent fevered experimentation guided the hand of nature into over 20,000 "cultivars," which despite their diversity are hybrids of only a handful of original species.

What is a species? Biologists originally preferred a definition based on morphology or appearance. A redwood looks like a redwood and a lion looks like a lion. Appearance was the basis of the first classification

Figure 3.1. An image of 50 different microscopic species of diatoms. Diatoms are at the base of many marine and aquatic food chains. In these examples, the morphologies are so distinct and distinctive that the traditional Linnaean method for classifying species works well.

system of biology in the mid-eighteenth century, by Carl Linnaeus. It generally works very well (Figure 3.1).[1] Ernst Mayr identified 137 species of birds in the mountains of New Guinea and botanists classified hundreds of plant species in the Chiapas region of Mexico, and in both cases 99 percent of their classifications agreed with the "folk taxonomies" of the locals. When experts agree with common sense, both sides usually declare victory and go home.

However, exceptions to the "If it looks like a duck" method are varied and vexing. Biologists found many examples of individuals within two populations that were hard to tell apart yet would not breed with each other, suggesting they were different species. A good example is two species of gray tree frogs that live in the central and eastern United States. They look the same, live in the same forest habitats, and eat the same kind of prey. During the mating season, however, they have unique calls that only attract members of the same species. The fact that they're different species is corroborated by differences in their number of chromosomes.

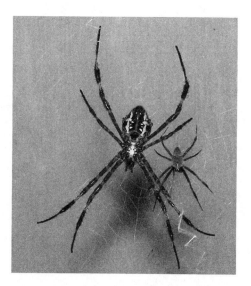

Figure 3.2. Extreme sexual dimorphism is common among spiders. The small male and larger female *Argiope appensa* seen here coexist uneasily, given that the male is likely to become lunch after sex. The size and morphology variations across gender can easily be confused as different species.

Morphology doesn't help much when it comes to classifying microbes such as bacteria, which reproduce by cloning.[2] Asexual reproduction also complicates the definition of species, such as when plants and some animals spread by budding. In addition, members of a species may look very different at different stages of their life cycle, such as caterpillars and butterflies. Or they may diverge due to mimicry—female butterflies of one species often change their appearance to match the local inedible species as a defense strategy.

The general environment can be responsible for differences that may make the casual observer think they're looking at different species. One species of arrowhead plant has distinctive arrow-shaped leaves when it grows on land, but long and spindly leaves when it grows in water. The hydrangea plant has flowers—more accurately, modified leaves—that can change from blue to red according to the pH of the soil and the amount of aluminum taken up by the plant.

A drab brown duck and a gaudy green, white, and black duck look to the untutored eye like members of different species, not female and male of the same species. Sexual dimorphism can be quite extreme.

Darwin was the first to recognize that the tiny creature found clinging to female barnacles is in fact the male of the species. Male spiders are often much smaller than their female counterparts, and we all know how that can end (Figure 3.2). Female anglerfish can be 100 times heavier than males. The male anglerfish is rudimentary and has no digestive system; it fuses with the female and only lives parasitically, becoming a sperm-producing system. Marine worms include a species where the male is hardly ever seen because it's tiny and lives in the female's genital sac. Divorce is not an option.

Species and Sex

If looks can be deceiving, what's the alternative? Since the 1940s the most widely accepted criterion has been the biological definition: a set of actually or potentially interbreeding organisms. A person can breed with another person but not with another species. At the genetic level the analogous concept is a gene pool: all the genes in an interbreeding population. All members of the species exchange genes with the pool, which defines the genetic diversity of a species. It seems simple, but once again the devil is in the details and the exceptions.

The standard biological definition can't be applied to any organism that relies most or all of the time on asexual reproduction. Among animals, certain lizard and salamander populations are all female. They lay eggs that have never been fertilized and that hatch as females to carry on the cycle. Asexual reproduction in plants is even more common, with dandelions and strawberries and redwoods as familiar examples. About half of the roughly 250,000 plant species, and the vast majority of flowering plants and ferns, reproduce by a process called polyploidy, where multiple sets of chromosomes are formed. We have one set of chromosomes from each parent, but plants can have three (banana, apple, ginger), four (potato, cotton, leek), six (wheat, oat, kiwi), or even eight (strawberry, pansy) sets.

Then there's the problem of hybrids. Carrion crows and hooded crows look different and largely mate within their own groups but sometimes they form hybrids. In other cases hybridization is confined to a zone.[3] Bullock's orioles and Baltimore orioles have their own areas, but in a vertical strip of the central U.S. they mingle and interbreed. The two populations are genetically distinct, even though genes are swapping across the boundary zone. Gene flow doesn't seem to homogenize the populations. Looked at in another way, hybrids tell us something about how different species actually form.

Breeding two different mammals that look broadly similar usually fails because the result is sterile. Mules are a good example. But mammal hybrids have been created in captivity, like ligers (lion-tiger), zorses (zebra-horse), wholphins (whale-dolphin), lepjags (leopard-jaguar), camas (camel-llama), and beefalos (bison-cow). It can happen on rare occasions in the wild too, as in the case of the blynx (bobcat-lynx), the pizzly (polar-grizzly), and the hybrid of gray wolf and coyote that led to the current red wolf thousands of years ago.

We're inching toward delicate terrain: the inviolability of the barrier between us and the other species.[4] If an animal is impregnated with human sperm, nothing happens; the species barrier is too great. We share 99 percent of our DNA coding sequences with chimpanzees, so how plausible is a humanzee? We have 23 pairs of chromosomes, chimps have 24; the party line is that the mismatch of chromosomes stops a viable fetus from forming. But in a close parallel, the domestic horse, with 32 chromosome pairs, was mated with Przewalski's horse, which has 33 pairs, to give semifertile offspring. Soviet scientist Ilya Ivanov attempted to cross humans with apes as part of Stalin's plan to create a superhuman soldier. There's no evidence that he succeeded, but the thought of the experiment gives most people the shivers.

Deciding whether two species can successfully breed is fairly simple, but applying the biological definition generally leads to hypothetical situations. We'll never know if the extinct creatures represented by two contemporaneous fossils could have reproduced. Then there are species that have a long lineage in the fossil record: could a trilobite from 340 million years ago have bred with a trilobite from 310 million years ago? What about the tricky phrase "potentially interbreeding"? Suppose two groups of snails are separated by a freeway. Crossing this barrier is fatal. We know the two groups *don't* interbreed, but it may not be enough to declare them separate species since we don't know whether or not they *could*.

On the topic of defining species, Darwin was not very encouraging. In 1856, he wrote, "It is really laughable to see what different ideas are in naturalists' minds when they speak of species. . . . It all comes, I believe, from trying to define the indefinable." The practical definitions mostly work, but the suppleness of biological mechanisms prevents scientists from getting overconfident.[5]

The definition of species is personal. When we say we're different from the apes, we're declaring separateness and, implicitly, superiority. Guy the Gorilla was the most famous inhabitant of the zoo when I lived in London as a child. A Western Lowland silverback, he weighed over 270 kilograms, his arm span was nearly three meters, and the girth of his neck was the same as a man's waist. He gently cradled small birds that flew into his cage, yet the fingers that could pick up a penny could have broken the neck of his keeper if he'd been so inclined. Now, I know as a scientist that his DNA almost perfectly overlapped mine. Then, I experienced him as a child, without preconceived notions. I remember feeling bad about the cramped and shabby squalor of his cage. And I remember staring into his eyes—the experience was visceral, almost shocking. I saw curiosity, I saw pathos, and I saw dignity. I saw a person.[6]

The Motor of Evolution

If we want to discuss the end of humanity, we have to know where we and the other millions of species come from. A hallmark of evolution is the formation of new species. It's remarkable—even to biologists who think about it all the time—that a universal common microbial ancestor led to organisms as diverse as fungi, algae, butterflies, elephants, and seahorses. Yet phylogenetic techniques can trace the broad shape of that evolution over 4 billion years by using the gradual divergence of the base-pair sequences of RNA and DNA. In a microbial world, plants and animals are twigs on the tree of life. In terms of the intertwined double staircase that is the blueprint of life, we're all one thing.

As life and lively conversation swirled around me in a local pub one evening a few years ago, I mused on the unity of life. Measured by accumulated deviations of the amino acid sequence for the protein cytochrome c, humans overlap 99 percent with rhesus monkeys, 84 percent with chickens, 68 percent with moths, and 60 percent with simple yeast.[7] I stared into my glass and realized we were not so different, the beer and I. If this was the diversity resulting from the rearrangement of the letters of one genetic alphabet, imagine what tales might be told with a range of genetic alphabets?

Every living organism is the physical manifestation of internally coded, heritable information. In biology, the genotype is the "blueprint" or set of instructions for building and operating an organism, written in the four-letter base-pair alphabet of DNA. The phenotype is the result, the physical manifestation of the organism. Biologists say that a genotype codes for a phenotype. The gene is the unit of information that codes for a particular trait and the sum of all genes in a population is called the gene pool. Mutation results in slight variation in the molecular form of a gene, called an allele. Then reproduction shuffles the alleles into new combinations in the offspring, which leads to variation in the phenotype.

Modern biologists have recognized that the "grain" of DNA and its microscopic variation don't explain the richness of evolution. They have embraced a new discipline called evolutionary developmental biology, or "evo-devo." Evo-devo recognizes that plant and animals are modular, often built on repeating parts like ribs or body segments. It also recognizes that gene control is complex—not only can genes be switched on and off, but they can also be regulated by the environment, and the same genes can be used differently by different organisms. Last, the ability of a genotype to change its phenotype in response to the environment means that genetic change can follow and not lead the development of novel features. Not all evolution is heritable.

The potential variation of our genetic material is phenomenal—10^{600} combinations of human alleles. You can be pretty sure nobody with your genetic makeup has ever lived or ever will! The environmental variation in the phenotype means that identical twins are never truly identical. Parents tell them apart and they have different fingerprints. With so much genetic possibility and so many ways to express it in a complex environment, there's plenty of raw material for evolution to act through natural selection.

All this gene shuffling and variation creates possibility as opposed to intention. Not everything has a meaning. Nature is parsimonious but sometimes it's lazy. Men have nipples because women do and it costs little to grow a nipple. Some attributes are vestiges but they are only mildly disadvantageous, like appendixes and wisdom teeth, so nature hasn't gotten around to getting rid of them. The sculpting of populations by the changing environment is the most powerful force in evolution and it creates more losers than winners.

The Evolution of Us

With that as background, how do new species evolve? We're trying to connect microevolution—the changes in the genetic composition of a population with the passage of each generation—with macroevolution—the appearance of a new, physically distinct life-form. Biologists like to argue a lot about the rate of evolution but it must be gradual; birds don't suddenly start flying and primates don't suddenly develop math skills. In a continuous span of evolution, deciding when a new species has emerged is difficult, which is why the practical biological definition is used wherever possible.

The formation of new species is governed by processes that push in opposite directions. Gene flow caused by reproduction tends to keep populations similar to each other, inhibiting new species. Populations will diverge due to natural selection and due to random variations in the frequency of alleles, especially in small populations, a process called genetic drift. The latter two mechanisms make the formation of new species likely. If gene flow is interrupted for any reason, genetic divergence ensures new species.

The most obvious way populations could be unable to exchange genes is geographical isolation. It doesn't take an ocean or a mountain range to isolate populations; it could be a variation in habitat in a contiguous region. Until 150 years ago, lions lived in open grassland in India and tigers lived in Indian forests. The species did not hybridize in the wild. Even when they live in exactly the same place there are other reasons species might not be able to interbreed. Just consider the problems of sex. Potential mates might breed at different times, they might meet and not be able to figure out the mechanics, they might not be able to transfer sperm, or they might not succeed in fertilizing an egg. Even if the egg is fertilized, the embryo may die, or the first generation hybrid might have low fitness, or be sterile. It's stressful to think about, which becomes yet another reason sex might fail.

All of these ideas are in play in the evolution of our distant ancestors. Mammals featured prominently in a burst of new species after the K-T impact 65 million years ago. About 35 million years ago, it's believed that some early anthropoid monkeys drifted on a raft of vegetation across the (then) much narrower Atlantic Ocean from Africa to South America. The New World and Old World monkeys were reproductively isolated thereafter and followed their own evolutionary paths after a surge of evolution (Figure 3.3).

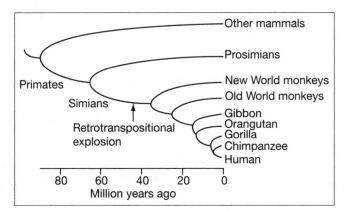

Figure 3.3. The evolutionary tree of our primate ancestors, showing a surge in diversity about 35 million years ago, which is inferred to have been caused by retrotransposons, a mechanism where genes relatively quickly duplicate and can create genetic novelty. Our last evolutionary connection with other mammals predates the K-T extinction event.

Old World monkeys split off from apes about 25 million years ago but coexisted in the same forest environments in Africa. In the competition for resources, there was a collapse of ape diversity; they couldn't keep up with monkeys whose shearing molars and gut specializations gave them a broader dietary range. In the cold, variable climate of the mid-to-late Miocene, 10 to 15 million years ago, the ability to exploit new sources of food was a decisive advantage.

The pivotal event in human evolution was the split between hominid and chimpanzee lineages about 7 million years ago, as revealed by fossil skulls. It was a shock when, in 2006, researchers showed that chimps and humans shared genes as recently as 5 or 6 million years ago.[8] In other words, we bred with chimps for over a million years before the second and final split. Victorians would not have approved and evolution is definitely not for the squeamish. Hybrid populations often go extinct because the males are sterile. The best guess as to what happened is that hybrid females mated with male chimps to produce viable offspring, and the human lineage emerged from the hybrid population.

Chimps have lived off fruit in tropical forests for millions of years, spinning off a number of variations, like gorillas who live on plants and hominids who moved into the drier woodlands and plains that opened up between the forests. From their branching point 5 or so million years ago, humans and chimps continued to diverge. In the sparser vegetation of the savannah, humans evolved bipedalism to cover larger distances and this gave them access to more types of food. They soon ranged over large sectors of southern Africa.

Around 3 million years ago the climate began to cool and undergo violent oscillations, with temperatures 4°C to 6°C (8°F to 10°F) cooler than at any time for millions of years. All species were stressed and some went extinct. Humans adapted by developing tools, by migrating, and by becoming hunters. *Homo habilis* developed a brain twice the size of any ape and geneticists have identified a single gene that may have facilitated the increase.[9] After spreading to the edge of the encroaching ice pack to find food, humans were the hunted as well as the hunters. This predation led to social cooperation that made explicit use of the newly larger brain. Chimps, by contrast, never come to the aid of each other, even when there is no cost to them. Social evolution increased the separation between the two species.

In recent human evolution, there have been instances of reproductive isolation. Australian aborigines were mostly cut off from other humans for 50,000 years. Yet when modern travel allowed them to meet other populations, they bred, as we know from the large number of common descendants today. Yet evolution doesn't stand still. There's evidence that a gene called *microcephalin*,[10] which regulates brain size, spread under selective pressure as recently as 37,000 years ago, just before humans began their rapid ascent to culture and civilization. This discovery shows the remarkable plasticity of the brain.

In the big picture, genetic changes over the past 150,000 years have been modest. If someone from that time were washed and shaved and dressed and dropped into your home or workplace, he'd fit right in. Just don't get in a fight with him over a taxi.

End of the Line

If success is measured by numbers, then 6 billion humans only sounds like a lot. Antarctic krill are a crucial part of the cold-water ecosystem and there are about 5×10^{14} of them. Even though each only weighs a gram, if they were put in a pile they'd weigh a billion tons, more than all humans. The most successful species is also an ocean dweller: the bacterium called SAR-11. This organism is also a key participant in the global ecosystem, acting as a "pump" to remove carbon dioxide from the atmosphere. There are 3×10^{29} of these cells floating around, give or take a few. That's 100 billion billion for each one of us.

If success is measured in terms of species longevity and diversity, the current champs are ants; they've been going for 100 million years and are represented by 20,000 species. Historically, hats off to trilobites, whose 17,000 species, ranging in size from microscopic to over a foot long, thrived for 250 million years from the Cambrian explosion to the Permian extinction. The 10,000 species of ammonites did even better, ranging from a few millimeters to three meters in size and lasting from 400 million years ago until the K-T extinction. Humans are newbies, with only a million years on the clock.

Earth has seen a half billion species in 4 billion years, and only 2 percent of those are alive now. The general rate of disappearance is called the background extinction rate. Species can be lost due to changes in the physical or the biological environment, and many factors are involved such as the population size, reproductive ability, genetic attributes, geographical distribution, and relationships with other species.[11]

On average, single-celled organisms and plant species last 10 to 30 million years and insect species last several million years. Among mammal species the mean is a million years. Biodiversity has been increasing strongly since the Permian "event" 250 million years ago, which means new species appear faster than others disappear. The cradle of current biodiversity is the tropics and the reason is a lower turnover at low latitudes—species in warm climates appear at a lower rate but they also disappear at a lower rate, meaning there are more around at any given time. Researchers have recently discovered that the energy required to create a new species is a fixed quantity.

In the past 500 million years, during which time the fossil record is good enough to count species and measure biodiversity, the steadily declining extinction rate has been superimposed by sharp spikes in extinction due to natural disasters. There are five "mass" extinction events—435, 370, 250, 205, and 65 million years ago—plus a similar number of more modest extinctions. About 95 percent of all land and sea species were lost at the end of the Paleozoic era, 250 million years ago. Most of these extinctions were caused by dramatic episodes of volcanism that altered the climate and the atmosphere, but impacts are also implicated, convincingly in the case of the K-T extinction 65 million years ago (Figure 3.4).

According to all the evidence, we're currently experiencing a sixth mass extinction or "Great Dying." In the fossil record the background loss of species is about 30 per year. According to the Millennium Ecosystem Assessment, commissioned by the UN, the current rate is 1000 times higher or 30,000 species per year. The projected future rate based on models of habitat destruction and global warming is 10 times higher

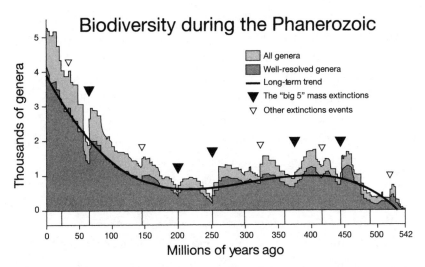

Figure 3.4. The evolution of biodiversity in the past half billion years shows no strong rise until 100 to 150 million years ago, when diversity in the oceans was matched by diversity on land. "Well-resolved genera" means groupings where the time resolution in the fossil record is good. The smooth trend is superimposed by five mass extinctions and a similar number of minor extinction events.

still. By some estimates, 30 percent of the world's plants and animals could be on a path to extinction within a hundred years. Unlike all previous extinctions, this one has our fingerprints all over it, which raises the question, If we cause an ecological disaster, will we survive it?

Our Own Worst Enemy

Dealing with our Effluent

Visualize Earth suspended in space, hanging like a pale pearl on the blackness of night. Its slender sheath of atmosphere is no thicker than the skin on an apple. Now zoom in and listen as the planet spins and dawn light streaks across the surface. Normally the sunrise would trigger a rising chorus of sound as birdsong surfs the gathering light at hundreds of miles per hour. But on this day there's only silence.

It's the mental image conjured up by *Silent Spring*, Rachel Carson's pioneering book from 1962, which helped launch the environmental movement. Her clarion call against uncontrolled pesticide use touched a chord, but it also spawned an ugly backlash. The chemical industry mobilized to deride her and discredit her and their spokesman Robert White-Stevens intoned, "If man were to follow the teachings of Miss Carson, we would return to the *Dark Ages,* and *insects* and *diseases* and *vermin* would once again inherit the Earth." But her claims were fully vindicated by a presidential advisory committee, and legislation that strengthened control of chemical pesticides soon followed. Our sense of Earth as a fragile ecosystem was cemented in 1968 when *Apollo 8* sent back the first images of the planet from 400,000 kilometers away.

When we were hunter-gatherers and only 10 million strong, we had a light footprint on the planet. We consumed only what we needed, left only biodegradable waste, and leaked a little heat and carbon dioxide into the atmosphere. Our lives depended on being able to hear and heed the planet's heartbeat.

Now there are over 6 billion of us, and we're addicted to growth and technology. We each generate nearly a metric ton of waste a year, less than a quarter of which is recycled. In the U.S. the discarded trash includes 50 million computers, 100 million cell phones, and 3 billion batteries a year. This rapidly growing electronic detritus contains lead, cadmium, chromium, mercury, and polyvinyl chlorides, which have toxicological effects ranging from brain damage and kidney disease to mutations and cancers. Not as toxic, but a headache for the planet nonetheless, are the 3,600,000 metric tons of junk mail, 22 billion plastic bottles, and 65 billion soda cans discarded a year. Americans are the world champion wasters, with 5 percent of the population but 40 percent of the trash.[12]

When we get rid of our trash, it doesn't totally go away. In a landfill some of the toxins leech into the groundwater and in an incinerator some escape into the air. The U.S. Environmental Protection Agency keeps track of 1300 of the worst toxic-waste sites, but the so-called Superfund that pays for cleanup has been bankrupt for years despite the intention from the original legislation that the polluter should pay. At the nation's 120 nuclear-waste sites, the decontamination problem is huge. There are 245 million metric tons of mine tailings from uranium ore, 45,000 metric tons of highly radioactive spent fuel from commercial and defense reactors, and over 340 million liters of high level waste left over from plutonium processing. Yucca Mountain in Nevada was never designed to handle this amount of waste.

Devra Davis is an epidemiologist who has consulted with the World Health Organization and advised every president since Jimmy Carter on public health policy. She has faced criticism from vested interests, just as Rachel Carson did before her, when she points out the many ways that the profit motive trumps the public good when it comes to avoidable hazards. She was just a baby when toxic smog fueled by a local zinc smelting factory rolled into Donora, Pennsylvania, killing 20 and making nearly half the town's residents sick. The incident helped spur the Clean Air Act of 1970.

Davis points out that we're blithely ignoring or being falsely reassured about the carcinogenic properties of ingredients in soda and makeup and over-the-counter drugs. In the U.S., one in two men and one in three women will get cancer so it's not just an individual killer, it's a population killer. Longevity, especially in developing countries, will depend increasingly on being smart with effluent. Just ask an Apollo astronaut what happens when you foul your spacesuit.

A Warm Embrace

Front and center on our plate of troubles is climate change. Over the past decade it's become clear beyond a reasonable doubt that human activity is heating up the planet. The primary causes are the excessive use of fossil fuel and economic development leading to deforestation. The 2007 report of the Intergovernmental Panel on Climate Change has been endorsed by 30 of the most prestigious scientific societies; the few naysayers that remain are flying in the face of massive evidence and unprecedented scientific consensus.

If you were sitting in your living room reading a book and I nudged up the thermostat 0.75°C (1.4°F)—the average amount of global warming over the past century—you probably wouldn't notice. So why is the world going to hell in a handbasket because of this effect? The problem isn't the warming so far; it's the effect of what's projected. This century should see another 1°C (2°F) to 5°C (9°F) of heating, where the variation is the result of the complexity of global heating and cooling mechanisms (Figure 3.5). The climate is like a supertanker; it can't turn on a dime. Even if we were to stabilize greenhouse gas emissions tomorrow, the warming would probably continue for centuries due to the heat capacity of the oceans. That leads to the concern that the changes are irreversible and we've passed a "tipping point."

By the measuring stick of nuclear war, or even the ensuing nuclear

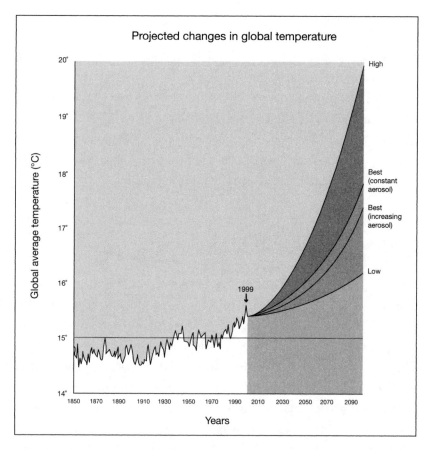

Figure 3.5. Assessment and projection of global warming by the Intergovernmental Panel on Climate Change, with various models projecting a range of 1°C (2°F) to 5°C (9°F) of warming by the year 2100. The range in model predictions relates to aerosol concentrations in the upper atmosphere and sensitivity of greenhouse warming to carbon dioxide and aerosol concentrations.

winter, global warming is a serious problem but not a crisis for the species. The working group for the intergovernmental report lists the impacts: wet areas will get wetter, dry areas will get drier, and the weather will get more extreme everywhere. Droughts and floods will become more common. The resilience of many ecosystems will be stretched to the breaking point. Carbon removal by living systems is expected to peak midcentury and then decline, which will amplify climate change.

Initially there'll be winners and losers. If you live in Bangor or Minsk it's hard to complain about a few degrees sliced off your winter. But early losers in terms of drought and loss of traditional crops include poor countries in Africa and Asia that can ill-afford impediments to growth. By 2050, almost all countries can expect to see ill effects.

Not to be parochial or trivialize the effects of global warming, but here's how it will affect things that may be near and dear to your heart. Get ready to wean yourself off lobster and salmon; they'll become scarce in the warming oceans. And you won't be drinking French wine or pinot noir with those meals, as the grapes will move elsewhere (hello, Brighton Beaujolais?). The classic ash baseball bat will become a rarity, as will Christmas trees. Fly-fishing streams and ski resorts will lose their reason for attracting paying customers. The only place you'll see polar bears, arctic foxes, and koalas is in zoos.

On the other hand, the warming will boost things we could do without. Burgeoning dandelions will ruin your lawn and rampant poison ivy, along with many more of the ticks that carry Lyme disease, will spoil your walk in the woods. Jellyfish and their painful effects will invade more coastal waters. Mold and ragweed will prosper with all the extra CO_2, so asthma and allergies triggered by the environment will be on the rise. Already, we've seen cholera hit South America, malaria strike Moscow, and the West Nile virus kill in Canada. The spread and resurgence of disease will continue as mosquitoes, mice, ticks, rats, and other carriers expand their range.

Fire in the Sky

The clock face is simple and stylized. One quadrant is shown because no more is needed; for the past sixty years both hands have hovered in view. It's the Doomsday Clock of the *Bulletin of the Atomic Scientists*— the iconic image of the nuclear age.

When Einstein derived the equally iconic equation $E = mc^2$ a century ago he understood it unleashed terrible possibility as well as amazing opportunity. Nuclear weapons release energy with millions of times more efficiency than the chemical energy behind all other types of explosion. In the depths of the cold war, after the Soviet Union and the U.S. escalated weapon development from the fission A-bomb to the fusion H-bomb, the Doomsday Clock inched forward to two minutes to midnight, a hairsbreadth from nuclear holocaust. Tensions eased with the Partial Test Ban Treaty in 1963, and the Anti-Ballistic Missile Treaty of 1972, but by 1984 the clock stood at three minutes to midnight. Dialog between the superpowers had ceased and they had amassed a staggering 70,000 nuclear weapons between them.

Since the end of the cold war, the threat of global nuclear destruction has been far from people's minds. The Doomsday Clock eased back to 17 minutes from midnight in 1991 with the fall of the Berlin Wall and the deep cuts to the arsenals of strategic weapons of both ideological adversaries. Yet in the past decade the clock has been inching closer to midnight again, and in 2007 it was moved from seven to just five minutes from the figurative apocalypse. Why?

In part, it was because the clock embodies other threats to humanity, such as global warming and biological weapons. But there were other factors. One was the growth of the nuclear club to 10, with potential future growth to 15 or 20, most of which are undemocratic countries. There's the related concern over 1500 metric tons of highly enriched uranium and 500 metric tons of plutonium, much of which is in unguarded civilian sites. But mostly it was the recognition that arms reductions talks have stalled, China's modest arsenal continues to grow, and the United States and Russia still possess 26,000 nuclear weapons, 2000 of which are already targeted and could be armed and launched in minutes.[13]

Moving the Doomsday Clock isn't done lightly. The *Bulletin's* Board of Directors consults with an advisory group that includes prominent policy makers and 18 Nobel laureates. Stephen Hawking, one of the board's sponsors, said in their 2007 press release, "As citizens of the world, we have a duty to alert the public to the unnecessary risks that we live with every day, and to the perils we foresee if governments and societies do not take action now to render their nuclear weapons obsolete." Even though it's 10 times smaller than its historical high, the current U.S. stockpile is equal to 140,000 Hiroshima blasts, or 1.8 trillion kilograms of TNT.

The loss of life in a large-scale nuclear war is difficult to estimate but it would be tens or maybe hundreds of millions. Humanity would be able to survive the bombs and the radiation they produced, but the chaos and instability that resulted might send civilization into decline. Even if we judge that a widespread nuclear war is very unlikely, more limited and highly plausible regional conflicts could have devastating effects due to the phenomenon called nuclear winter.

Nuclear winter describes the climatic effects of a nuclear war, where fires initiated by the weapons deposit large amounts of smoke and soot in the upper atmosphere, where it causes sharp and long-term cooling, with a resulting famine for people far from the target zones. It entered the public consciousness in the early 1980s thanks to Carl Sagan and his public statements regarding a paper he co-authored on the subject. In an interview in 2000,[14] Mikhail Gorbachev said, "Models made by Russian and American scientists showed that a nuclear war would result in a nuclear winter that would be extremely destructive to all life on Earth; the knowledge of that was a great stimulus to us, to people of honor and morality, to act in that situation."

The original nuclear winter research was controversial, and in some respects the predicted effects were exaggerated due to limitations of computer modeling in those days. Recent research recasts the threat in a sobering way. One study presented at the American Geophysical

Union in 2006 found that a small-scale, regional nuclear conflict could result in as many fatalities as World War II and enough soot sent into the atmosphere to cool the grain-growing areas of the world by a few degrees, with a "catastrophic" effect on food production. In 2008, the *Proceedings of the National Academy of Sciences* published another study that looked at a similar scenario and found the extra effect of loss of 50 percent of the ozone cover at northern latitudes.

Both studies assumed subtropical countries using 50 warheads each. The countries are unnamed but India and Pakistan are each estimated to possess 70 to 80 warheads and they've been involved in three major conflicts in the past 60 years. Taking the larger view, we can hope that we're emerging from our troubled adolescence as a race with the ability to destroy ourselves many times over, and that the stakes are high enough that we'll all keep our eyes on the nuclear ball.

Biological Terror

Biological terror dates to Roman times, when dead or rotting animals were thrown into wells to poison water supplies. In the Middle Ages, some armies even used bubonic plague to lay siege to cities, but the aggressors were often unable to control their own weapons. Mustard gas was used by both sides in World War I, and revulsion over its use spurred the Geneva Protocol of 1925. A succeeding treaty, called the Biological Weapons Convention, has been signed by an impressive 185 countries, but the nonsigners include ominous names: Angola, Egypt, Iraq, North Korea, Somalia, and Syria. Biological weapons have been used by dozens of countries during wars and on occasion against their own citizens. About 40,000 metric tons of chemical weapons are stockpiled in hundreds of secure and sort-of-secure locations around the world.

In the public mind, bioterrorism is most strongly associated with the only two incidents where civilians have died at the hands of civilians. In 1995, a terrorist group called Aum Shinrikyo released sarin in the Toyko subway, killing 12 and making over 5000 people sick. In

2001, letters containing anthrax spores were sent to targets in Washington, DC, killing 5 and infecting 17 others. The anthrax-laced letters were purportedly sent by government bio-defense research scientist Bruce Ivins, who subsequently committed suicide.

Biological weapons fall into three categories. Nerve agents, like sarin, are artificial creations that attack the nervous system by interrupting the breakdown of the neurotransmitters that cause muscles to relax. Nerve agents can cause neurological damage and death, but they're not contagious. Biological agents include entities as diverse as prions, fungi, and parasites, but the two common types are viruses— such as those that cause yellow, Dengue, and Ebola fevers, and smallpox—and bacteria—like those causing anthrax, plague, cholera, and botulism. A biological weapon's effectiveness is governed by its lethality and mode of transmission. Those that can be spread by contact or as an airborne aerosol are the most worrisome, especially if they have a latency of days in which case the disease will already have been transmitted to others by the time it's recognized.

Our defenses against these pathogens are minimal. Viral hemorrhagic fevers like Ebola and Marburg are 30 to 80 percent lethal and have no effective treatment. There are vaccines for anthrax, botulism, plague, and other bacterial agents, but they're not available to the public and couldn't be delivered in sufficient quantity to the location of an attack. The specter of bioterror centers on genetically modified agents, for which vaccines don't exist.

To get a sense of the scale of the problem, let's take a glimpse into the Soviet biological weapons program. The Soviets used a bacterial agent called Tularemia during the 1944 Stalingrad campaign against German troops, afflicting over 100,000. They continued to develop and mass-produce biological weapons even after signing the Biological Weapons Convention in 1972. Spores of weaponized anthrax were accidentally released near Sverdlovsk in 1979, killing at least 100, though nobody knows the exact toll since the KGB destroyed all evidence and records. By the 1980s the Soviets had established Biopreparat, a vast network of 20 secret labs, each one focused on a different deadly agent, with 30,000 employees, including thousands of PhD scientists. The effort shrank with the fall of the Soviet Union but is still substantial. Western governments have little idea of what goes on in the still-secret labs.

What little we know is worrisome enough. For 20 years, Sergei Popov was one of the top scientists at a Siberian research facility that worked on genetically altered pathogens. In Project Bonfire, he helped create plague bacteria that were resistant to 10 different antibiotics, and anthrax modified to resist *all* vaccines. In the very scary Hunter Program, whole genomes of viruses were combined to produce hybrid and untreatable viruses. Interviewed for a 2001 *Nova* special, he said, "Imagine a bacterial agent which contains inside its cells a virus. The virus stays silent until the bacterial cells get treated. So, if the bacterial disease gets recognized and treated with an antibiotic, there would be a release of virus. Then after the initial bacterial disease was completely cured, there would be an outbreak of a viral disease on top of this. It could be encephalomyelitis. It could be smallpox. It could be Ebola. Those viruses were on the list of potential agents."

Another example is equally chilling. Popov and his researchers inserted mammal DNA into a bacterium responsible for a low-mortality form of pneumonia. That DNA expressed fragments of a protein called myelin, the electrically insulating fatty layer that sheathes our neurons. In test animals, the pneumonia came and went, but myelin fragments carried by the recombinant bacteria goaded the animals' immune systems into reading their own natural myelin as pathogenic and attacking it. Brain damage, paralysis, and 100 percent mortality were the result. Popov created a biological weapon that triggered rapid multiple sclerosis.[15] He claims the agent was not weaponized in large quantities, but the knowledge needed to do so is readily available.

Even if we don't trust Russian intentions, it seems farfetched that humanity could be at risk. But advances in genetic engineering have been breathtaking; what was done in the past by the Russians with great effort and vast expense can now be done with gene-sequencing equipment bought secondhand on eBay, using unregulated biological material that arrives in a FedEx package.

In 2008, a bipartisan congressional committee on prevention of the spread of WMD and terrorism raised the alarm, noting, "Our margin of safety is shrinking, not growing." They went on to say, "The biological threat is greater than the nuclear; the acquisition of deadly pathogens, and their weaponization and dissemination in aerosol form, would entail fewer technical hurdles than the theft or production of weapons-grade uranium or plutonium and its assembly into an improvised nuclear device."[16] It's no longer a chilling science fiction scenario that some malign group could build the perfect pathogen and unleash it on the world. Luckily, we can use the same genetic engineering tools to protect ourselves.

Even if we keep a lid on the man-made threats, evolution doesn't stand still and we might be tripped up by something beneath our attention. In the opening vignette we encountered a tiny misshapen protein that occurs naturally and has the power to erode brain tissue and send the victim into a downward spiral of tremors and dementia. It achieves this by inducing normal proteins in the host to adopt the dysfunctional configuration. We should be thankful the incidence of this disease is only one in a million because the prion's strategy is effective and its mode of propagation is lethal. We're fighting something that isn't even alive by a biologist's definition; prions couldn't exist if they killed off all their hosts, but they're too simple to care.

Bioterror and nuclear holocaust have spawned subgenres of apocalyptic thought. If you're looking for offsetting cheer, it comes in the form of all the other "end of the world" scenarios that flare up in the popular media like bouts of acne. The most recent one alludes to the Earth being roasted by solar flares or hit by a rogue planet called Nibiru in 2012. There's no scientific credibility to the threat, which is based on a misreading of Mayan mythology. From fear over the return of Comet Halley in 1910 to the mangling of Nostradamus's words into Y2K doom, these claims share one thing: they were dead wrong. So I publicly offer this to any person or group claiming knowledge of the end of the world: I'll bet all my assets against yours that you're wrong. This is the perfect sucker bet—if the end of the world doesn't come, I'm richer, and if it does, the bet will have been voided by fate.

Chapter 4

BEYOND NATURAL SELECTION

The squibbons pause in their conversation and watch thoughtfully from the treetops as the shagrats pass by underneath, noses twitching. The outsized rodents are fierce adversaries with septic bites, but they will not challenge a troupe of well-armed squibbons. The air is loud with the buzz of bumble-beetles foraging. A toraton passes by the edge of the forest looking for new growth to eat. Its footfall shakes the trees and its head can be seen above all but the highest treetops.

Welcome to the Earth of the future, after 200 million years of climate fluctuations and evolution have created legions of new winners and losers. British geologist Dougal Dixon ventures out of the box to envisage a world order that does not feature humans. The new "top dog" is a hybrid between squids that have returned to land and a descendant of small monkeys.

In addition to fist-sized flying beetles and house-sized reptiles, Dixon imagines a world inhabited by the "ocean phantom," a jellyfish the size of a truck, the "swampus," a jellyfish that emerges from swamps to feed, and the "megasquid," a multiton land-based squid. Not all the creatures are huge; "flish" represent a successful evolutionary pathway from fish to birds.

Geologist Peter Ward at the University of Washington takes a more sober approach to speculative biology, basing his predictions on current pressures. He believes global warming and the loss of grasslands will spell the end for large mammals such as lions and tigers and bears (oh my). The big winners will be creatures in the category known as "supertaxa," which diversify rapidly and have a relatively low extinction rate. Environmental stress will favor snakes, cockroaches, foraging birds, and—in this he agrees with Dixon and many evolutionary biologists—outsized rats.

These successful species drive each other to greater diversification, snakes eat rodents so they would respond to any new capability or advance among the rodents. Ward is more sanguine about the future of humans than Dixon but admits that the crystal ball is murky. He thinks our current brute force tinkering with evolution could have bad and unintended consequences.

Endpoints of Evolution

Comings and Goings

It's too early to tell whether the hairless ape with the dangerous toys will remove itself from the gene pool. Hopefully not, then we'll be a part of nature and not apart from nature. We're young as a species, just 200,000 years old. Our distant ancestors *Homo erectus* survived 1.5 million years; setting aside the cautionary tale of our Neanderthal cousins, who went extinct after 100,000 years, we'd like to think we're just getting going.

Evolution is driven by mutation and a changing environment, but the most successful species have great durability. Sharks and crocodiles have persisted for tens of millions of years with few changes because they're very well adapted. Some simpler organisms do even better. Anaerobic bacteria were abundant on early Earth and they still find happy niches in the modern environment. Bacterial colonies called stromatolites have lived in coastal reef ecosystems continuously for 3.5 billion years. Adaptation doesn't have to be perfect; it just has to be good enough. The vertebrate eye with its back-to-front wiring and blind spot is a good example. Life's chancy but if the game resembles 21 then some species have built a decent hand and are sitting tight.

Species come, and species go, and keeping track of them isn't trivial. About 2 million have been identified and cataloged and best guesses are that this is less than 10 percent of the total. The current rapid extinction rate is an additional problem. As biologist Daphne Fautin, a member of the international team that assembles lists of new species, said, "While we are finding many new species, a lot more are disappearing faster than we can discover them. Many are gone forever and we'll never know they existed."

Over half of known species are insects, including 350,000 types of beetle. When British geneticist J. B. S. Haldane was asked what his study of nature had told him about the Creator, this was his answer: "If He exists, He has an inordinate fondness for beetles." Academic paleontology is riddled with disagreements between "splitters" and "lumpers," those who classify organisms into more or fewer species, and "jerks" and "creeps," those who think evolution progresses in surges as opposed to smoothly and continuously.

We've probably discovered most of the mammals and the birds, but little is known about other groups, like nematodes. When it comes to tiny species, we've barely scratched the surface, so to speak. In one study, 10,000 bacterial species were found in one gram of Minnesota soil, less than a quarter of which had ever been cultured or classified, perfect rebuttal to those who claim there's not much life in the "Land of 10,000 Lakes."

The task of cataloging biodiversity is complicated by how much of it's in exotic or remote places; 70 percent of known species live in 12 countries: Australia, Brazil, China, Colombia, Ecuador, India, Indonesia, Mexico, Madagascar, Peru, and Zaire. It's not easy work for naturalists. Carl Linnaeus, the eighteenth-century scientist who is the Father of Taxonomy, sent many students out from Sweden with instructions to zealously search for new species. A third of them died on their travels. But the reward was to use his system to give a name to a new creature.

What's in a name? That which we call a rose by any other name would smell as sweet. If you ever discover an animal, vegetable, or mineral, tradition gives you the privilege of naming it. Normally it's done with judiciousness and decorum, but not always.

Linnaeus started the slippage by naming an ugly weed after a critic. Richard Fortey, senior paleontologist at New York's Natural History Museum, wrote about a colleague who hated Communists, so he named a worm he had discovered *Kruschevia Ridicula*, but he loved punk, so he named two trilobites *Sid viciousi* and *Johnny rotteni*. Quentin Wheeler named slime-mold-eating beetles after George W. Bush and members of his cabinet but swore that he was a lifelong Republican and that the action was a compliment. Scotsman G. W. Kirkaldy was more romantic, giving a set of bugs the same Greek suffix so that it became a list of his romantic conquests: *Florichisme, Marichisme, Peggichisme,* and so on.[1]

Sometimes there's a long story behind a naming but other times the reason is fairly obvious. Consider some new names from the past couple of decades: beetles called *Agra vation* and *Agra phobia*, a pine tree called *Pinus rigidus*, a mollusk called *Abra cadabra*, an extinct rat-kangaroo called *Wakiewakie,* a spider genus called *Orsonwelles*, and a genus of snails a bit smaller than *Bittium* snails, *Ittibittium*. You don't have to be a naturalist to play. In 2005, a gambling Web site shelled out $650,000 to name a Bolivian monkey and in 2007, a conservation group auctioned off the rights to name 10 fish species, with the money going to preserve their habitat. Stephen Colbert invited a biologist on his cable TV show and goaded the scientist into naming a spider after him.[2]

Naming things is reassuring but tracking their passage in and out of the gene pool is much harder. Incomplete sampling that doesn't give us a full view of biodiversity means we might declare a species to be extinct when it's not: the coelacanth, rediscovered in 1938 after being thought to be extinct for 80 million years, is the prime example of this Lazarus effect. Sometimes a species comes within a hair's breadth of erasure, yet endures.

Hanging by a Thread

With 6.8 billion of us, humans are far from being endangered. It was not always this way. The story of our ancestors is told in mitochondrial DNA and the Y chromosome, the two parts of the genome that aren't shuffled by evolution from generation to generation. In the "recently out of Africa" model, all of us inherited DNA from a mitochondrial Eve who lived in Africa about 160,000 years ago and all men inherited their Y chromosomes from an African who lived 60,000 years ago. Just after the species emerged it split into a southern group, who became the modern day Bushmen, and an eastern group, who became everyone else—including us.

Africa was racked from droughts between 90,000 and 135,000 years ago and the DNA evidence shows that the population went through a severe "bottleneck," perhaps dropping to as few as 2000 individuals. After the droughts the population recovered and there were about 40 groupings of humans scattered across Africa. Further environmental stress was caused by the eruption of a supervolcano called Toba in Sumatra 74,000 years ago, causing severe cooling for a number of years. About 60,000 years ago, two small groups, perhaps each only several hundred strong, started the epic migrations that populated the world. Humans spread to Asia first, then Australia 50,000 years ago, Europe 35,000 years ago, and the Americas 15,000 years ago.

When genetic diversity is reduced in a population bottleneck it can cause extinction because the survivors might not be able to adapt to new selection pressures. We were hanging on by a thread. But for us the story turned out well. Spencer Wells, director of the Genographic Project and explorer-in-residence at the National Geographic Society, said, "This new study illustrates the extraordinary power of genetics to reveal insights into some of the key events in our species' history. Tiny bands of early humans, forced apart by harsh environmental pressure, come back from the brink to reunite and populate the world. It's truly an epic drama, written in our DNA."[3]

The hardships that lead to a population bottleneck play both ways in subsequent survival. The environment is different so new behaviors and strategies have to be developed quickly. It's as if you'd gotten good at playing tennis all your life and suddenly the game was Roller Derby. But species that develop new behaviors have comparative advantages in the struggle for resources. In terms of genetics, bottlenecks are bad news. Genetic drift increases but the gene pool shrinks so maladaptive genes survive and propagate because they aren't culled as efficiently in a diminished population. Researchers have found that humans and chimps have each accumulated about 140,000 nonadvantageous DNA mutations since the lineages split. In the same interval rats and mice, whose genome is largely similar, have accumulated substantially less. That makes us more susceptible than rodents to genetic diseases like cancer (Figure 4.1).

How slender does the thread get before it breaks? In ecology, there is a minimum viable population where models predict a 90 to 95 percent survival probability 40 to 50 generations in the future. For large mammals, this number is about 50 individuals.

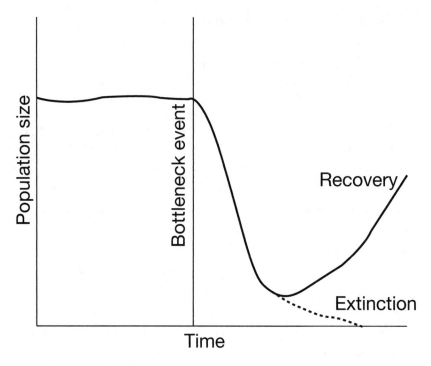

Figure 4.1. A population bottleneck is caused by virulent disease or an ecological disaster. When the population drops very low, extinction or slow recovery can follow. The diminished gene pool causes genetic drift, which is a spur to evolution but can also cause inbreeding and susceptibility to disease.

Of Microbes and Men

In the evolutionary arms race between microbes and men, the bugs are moving far ahead. As you read this, the 10 trillion cells in your body are outnumbered 10 times by bacteria, from nearly a thousand different species, many of which can't be cultured in the lab. Much of the microbial biomass is underground or in the oceans where it's not readily visible, and even though each one is only a quadrillionth of a gram when toweled dry, overall they weigh a gigaton and there are roughly 6×10^{30} of them.

Most are benign, but their malignant cousins—viruses and prions—are looking to cause trouble and our metaphorical doors and windows are wide open. Not only do we live in an overwhelmingly microbial

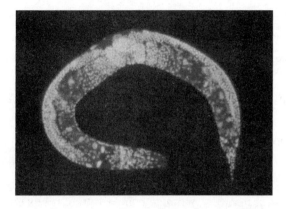

Figure 4.2. Much larger than a bacterium, this millimeter-long worm called *Caenorhabditis elegans* may hold a key to parasitism in humans. *C. elegans* is one among tens of thousands of nematode species, found in all of the ecosystems on Earth. The nematode cohabits with Oriental beetles in the U.S. and Japan so it can eat the bacteria and fungi that live on the beetle's carcasses. Parasitic nematodes like this infect 2 billion people worldwide and severely sicken 300 million.

world, but the bacteria contribute more genetic material as residents of our bodies than our bodies do. Moreover, even the concept of "them" and "us" is misleading. Bacteria share our food but they interact with our bodies and the environment in such complex ways that it's better to think of people as human-bacteria hybrids. Bacteria are the threads from which human cloth is made (Figure 4.2).

Coexisting with a complex microbial ecosystem is disconcerting, but it's the comparative rates of evolution that place us at a disadvantage. In 10 years bacteria produce 200,000 generations, the same number of generations humans have produced since splitting off from chimps. In less than a human generation we've seen the emergence of dozens of antibiotic-resistant bacteria and devastating new diseases such as AIDS—10 billion new viral particles are created each day in someone infected with HIV. When bacteria are exposed to antibiotics they must become resistant or die. Both viruses and bacteria accelerate evolution by transferring large chunks of DNA in a process called horizontal gene transfer. Humans are in an evolutionary quandary like that framed by the Red Queen in Lewis Carroll's *Through the Looking Glass*, where "It takes all the running you can do, to stay in the same place."

Listen to Joshua Lederberg, a Nobel Prize winner for his work on the genetics of bacteria, in a commentary from 2000 in *Science* magazine: "What makes microbial evolution particularly intriguing and worrisome is a combination of vast populations and intense fluctuations in those populations. It's a formula for top-speed evolution. Microbial populations may fluctuate by factors of 10 billion on a daily cycle as they move between hosts, or as they encounter antibiotics, antibodies, or other natural hazards. A simple comparison of the pace of evolution between microbes and their multicellular hosts suggest a million-fold or billion-fold advantage to the microbe. A year in the life of bacteria would easily match the span of mammalian evolution! By that metric, we would seem to be playing out of our evolutionary league."[4]

His last name is hard to live up to but Larry Brilliant has tried. He took an unusual path to becoming an internationally noted epidemiologist and philanthropist. It included being the doctor for Native Americans when they took over Alcatraz Island in 1969, being a Bollywood extra, cofounding one of the first online communities in 1985, and helping save the sight of 2 million people in developing countries. Probably his biggest achievement as a young physician was being a part of the team that eradicated smallpox.

Brilliant was appointed the executive director of Google.org in 2006. Google.org is the philanthropic arm of the search giant, and his work was backed by a commitment from Google's founders to devote 1 percent of the company equity plus 1 percent of the profits to good works. Brilliant wants to turn the Internet into a sentinel to warn the world of any impending pandemic, whether bird flu or a new variant of HIV. He says "We hope to develop an entire new science of epidemiology and surveillance both for existing diseases and to spot emerging ones early on . . . also using tabletop planning exercises with modern war-gaming techniques to better prepare a pandemic response."

How will it work? Software agents and bots will crunch Web reports on the incidence of symptoms like fevers or respiratory prob-

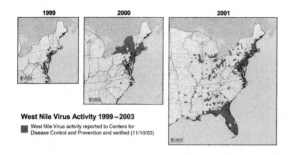

West Nile Virus Activity 1999–2003
■ West Nile Virus activity reported to Centers for
Disease Control and Prevention and verified (11/10/03)

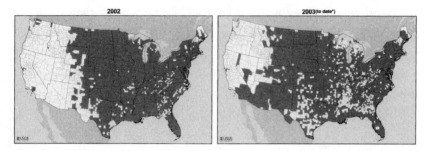

Figure 4.3. Maps of the spread of the West Nile virus in the United States from 1999 to 2003. The agent spreads quickly through population centers and then fills in geographically until it is pervasive. Pandemics often follow this pattern.

lems, Google Earth will be used to look for patterns or geographic trends of illness, and this will all be combined with genetic information on potential and existing pathogens.[5] "Had we been there at that very moment of the birth of HIV/AIDS," he says, "we would have been able to prevent this horrible pandemic."

In 2008, Brilliant moved on from Google to head the Skoll Foundation's Urgent Threats Fund. It's easy to see why he is so worried about pandemics. Beyond the increasing prevalence of antibiotic-resistant bacteria, there's an increasing occurrence of diseases that jump the species barrier from animals to humans. These include bird flu, rabies, SARS, Japanese Encephalitis virus, Dengue virus, West Nile virus, and Lassa fever (Figure 4.3). Viral fevers are probably the most frightening because they tend to be lethal and have no cure. Between 2000 and 2005, 50 million people caught diseases from animals and 100,000 died. We'll need a brilliant line of defense to hold back this tide.

The Future Us

If natural selection is defined by the phrase "survival of the fittest," we're in deep trouble. The rate of obesity hovers around 30 percent in most industrialized countries and the average adult is winded after running up a flight of stairs. Nearly 50 percent of Americans need vision correction, and 30 percent of births occur by C-section rather than naturally. The overprescription of drugs and a general paranoia about germs leads many adults to have frail immune systems with very little resilience. Shorn of technology and dropped into the situation of a hunter-gatherer 10,000 years ago, many of us wouldn't survive a month.

We've chosen instead to remove ourselves from the game. Tools and technology appear to have exempted humans from the hurly-burly of Darwin's heartless landscape. By design, we're adapting ourselves to the things we do the most—driving, watching TV, and slouching at a desk staring at a computer.

The conventional wisdom holds that human evolution has slowed and might have even ground to a halt. British geneticist Steve Jones has said, "Small populations which are isolated can evolve at random, as genes are accidentally lost. Worldwide, all populations are becoming connected and the opportunity for random change is dwindling. History is made in bed, but nowadays the beds are getting closer together. We are mixing into a global mass, and the future is brown."[6]

Jones also points to changes in reproductive patterns. A drop in the number of older fathers worldwide is having an effect on the mutation rate. Cell divisions increase with age and with each division there's a chance of an error or a mutation. There are 300 divisions between the sperm that made a 29-year-old father and the one he passes on. For a 50-year-old father the number is over a thousand. Jones also thinks our population growth implies that the power of natural selection is weakening. "In ancient times half our children would have died by the age of twenty," he says. "Now, in the western world, 98 percent of them are surviving to twenty-one." He argues that three major components of evolution—natural selection, mutation, and random change—are being reduced in humans.

Oliver Curry, evolutionary theorist at the London School of Economics, accepts this overall picture but adds the prediction of a divergence of the human line into two strands in the next few thousand years. With growing wealth and travel, cultural differences will be reduced. A tall, intelligent, coffee-colored·elite emerges, but sexual selection means they diverge from the majority, who because of overdependence on technology degrade into short, dim-witted, ugly, goblin-like creatures. According to Curry the outcome is "gracile" and "robust" humans like the Eloi and Morlocks from H. G. Wells's novel *The Time Machine*. Let's hope this dystopian scenario never emerges.

Yet as we learn more about the nuts and bolts of our genetic material, evidence accumulates that human evolution is more active than ever. Over the 6 million years since humans and chimpanzees split, our DNA has been changing seven times more rapidly than that of the chimps. While current intermixing of humans does keep any subgroup from diverging, population growth ensures increasing internal variation within the human gene pool, and cross-ethnic and international mating recombines our genes at unprecedented rates.

In 2005, Bruce Lahn of the University of Chicago identified two genes important in brain development that swept through the population. The *microcephalin* gene arose roughly 40,000 years ago and is carried by 70 percent of all people, and one variant of the *ASPM* gene arose roughly 10,000 years ago and is present in 25 percent of all people. Nobody knows what these genes are for, but other examples are clearly important for success and survival.

Lactase is an enzyme that allows people to drink milk their entire lives; it appeared after cows were domesticated. Another set of genes began to appear 10,000 years ago, as people settled in larger groups and experienced epidemics. Now they've given some fraction of the population resistance to malaria, smallpox, and even AIDS.

Anthropologist John Hawks has quantified the evolutionary "churn" by looking for alleles that correspond to the difference of one nucleotide between individual versions of the genome. He focused on places in the genome where genetic variations occur more often than can be explained by chance, usually because those changes give a selection advantage. The research group found evidence of recent selection on 1800 genes, or 7 percent of the genome.

Positive selection in the past 5000 years—the span of civilization—has occurred 100 times faster than in any other period of human evolution. "We are more genetically different from people living 5000 years ago than they were different from Neanderthals," according to Hawks. It would be fascinating to come back in 50,000 years and see what we've turned into.

Beyond Biology

More than Human

In the tenth tablet Gilgamesh is coming to terms with the death of his friend Enkidu. Part-man, part-God, and a powerful and restless king, Gilgamesh embarks on a perilous journey to visit the only two humans who were granted immortality by the gods, in the hope that he too can attain immortality. He travels by night through the land where the Sun is protected by scorpion-beings. He kills stone giants who accompany the ferryman only to find out that they are the only creatures who can cross the Waters of Death, which are not to be touched. So Gilgamesh cuts 300 trees into oars so he can discard each one after it enters the water. Finally, he reaches the island of Utnapishtim and his wife, and asks the immortals for their help. They chastise him, saying that fighting the fate of humans is futile and spoils the joy in life.

The Epic of Gilgamesh, originally called *He Who Saw the Deep*, is one of the oldest works of literary fiction. It is an epic poem from ancient Mesopotamia and it strongly influenced Homer's *Odyssey* and stories later included in the Bible. Gilgamesh may have been a true historical figure, the fifth king in the First Dynasty of Uruk in the third millennium BC. His story speaks to us across the centuries because it's about the anguish of loss and death. Gilgamesh was enough of a god to taste the benefits of power and longevity, but he couldn't make the transition to immortality (Figure 4.4).

Ray Kurzweil plans to do better. He takes 250 diet supplements daily, drinks only alkaline water and lots of green tea, and avoids high-risk activities. He has some reason for concern; his father and grandfather both died from heart disease. Kurzweil has type 2 diabetes, which he controls by his diet without taking insulin.

He sounds like a crank, but Kurzweil has an impeccable pedigree as a technologist and futurist. At age 8 he built a miniature theater where a robotic device moves the scenery, and at 16 he built a computer and programmed it to compose music. He's won the Lemelson-MIT Prize, which is the top award for inventors, and he's a National Medal of Technology winner. His core message ties to the current exponential advance

Figure 4.4. Gilgamesh, shown on a relief from the Assyrian capital Khorsabad, was a legendary hero in search of immortality. The story presaged many themes and leitmotifs from later religions, and in his yearning for the unobtainable, the half-god, half man Gilgamesh is an empathetic character.

of computation, nanotechnology, and genetic engineering. He thinks these rapid advances will converge in a time when humans will be able to transcend biology and live forever. This isn't a misty science fiction scenario; Kurzweil thinks the "singularity," as he calls it, will arrive in about 30 years.[7]

We're already marching swiftly down this road and the technology is outpacing our ability to control it or understand its social implications. Joel Garreau wrote his book *Radical Evolution* to draw the attention of a wider audience to the genetic revolution. Viagra, Botox, steroids, and SRIs for depression—these are just the first primitive manifestations of the ways we'll be able to alter ourselves using gene therapy. Modified humans will first appear on the sports field and the battlefield, but the same engineering of the genome will reach a wide audience as its cost goes down. Even so, Garreau worries about social stresses resulting from a group of people he calls "the enhanced," who can afford to live longer and become smarter and sexier, and "the naturals," who forgo genetic enhancement for philosophical or religious reasons. A third and majority group are people excluded from this revolution for financial reasons, who might be justifiably resentful.

In Kurzweil's vision, computation and medical technology will converge in a capability to repair and replace our bodies from within. He argues that a central trope of science fiction—man versus machine—is wrong. Instead, we'll meld with technology and *become* the machine. We'll have many millions of blood cell–sized robots, or nanobots, swarming through our bodies, patrolling for pathogens, and repairing our bones, muscles, arteries, and brain cells. Kurzweil says, "Death is a tragedy." These indefatigable repair crews will destroy disease, rebuild organs, and remove natural limits to our intelligence. Genetic improvements will be downloaded from the Internet. It's a classic Utopian vision.

To Heaven or Hell

The vision has a flip side. If technology can conquer aging and we can evolve into cyborgs, immortality beckons. But apart from the massive ethical question, should we? there's the practical question, what if it goes wrong?

The icon of the dystopic scenario is Bill Joy, whose tech credentials are as impressive as Kurzweil's. Joy was the primary author of the UNIX operating system, lead developer of the Java programming language, and he founded Sun Microsystems in 1982. Spurred by a conversation in a bar with Kurzweil, he authored a sobering and influential essay in *Wired* magazine called "Why the Future Doesn't Need Us." Joy thinks the dangers of these technologies are so great that we should have a moratorium on research in some areas, and let politicians and social scientists catch up in their understanding of the perils and pitfalls. His favorite example is the Australian researchers who were searching for a mouse contraceptive and started tinkering with the mousepox virus. It's harmless to humans but a close relative of smallpox, which is fatal to humans. With one small change to its genetic structure, the virus became 100 percent fatal to mice, a shocking result to most geneticists.

So much of the information on genetic engineering is on the Web, and the methods are getting so easy to acquire, that Joy thinks a disaster is almost inevitable. "If you handed a million people their own private atomic bomb," he asks, "do you suppose one of them would be crazy enough to use it?" He's worried that we have a chance of extinguishing the human race in the next 25 years. Joy warns, "Failing to understand the consequences of our inventions while we are still in the rapture of discovery and innovation seems to be a common fault of scientists and technologists; we have long been driven by the overarching desire to know that is the nature of science's quest, not stopping to notice that the progress to newer and more powerful technologies can take on a life of its own."[8]

Joy makes the simple point that a bomb can only be used once, while robots, engineered organisms, and nanobots might be able to replicate in the near future. In the view of roboticist Hans Moravec, this isn't a bad thing. He thinks we should proudly work to create robots that will supplant humans as Earth's superior race. They're our progeny too, he points out, and we should get out of the way of their ascendance.

Many people feel real discomfort in the face of views like this. Luckily, technology doesn't progress outside the social sphere. Over the history of nuclear weapons and nuclear power, and now with stem cells and cloning, there has always been a vigorous debate on the trade-offs of each technology, and society has usually acted to impose regulations and controls. There's no reason to believe this won't also happen with nanotechnology and genetic manipulation.

Even those suspicious of technology might feel the tug of the futurist vision, for example, as espoused by Gregory Paul and Earl Cox in their book *Beyond Humanity*: "We are dreaming a strange, waking dream; an inevitably brief interlude sandwiched between the long age of low-tech humanity on the one hand, and the age of humans transcended on the other. . . . We will find our niche on Earth crowded out by a better and more competitive organism. Yet this is not the end of humanity, only its physical existence as a biological life-form." But we are very unruly apes, and wisdom isn't our strong suit, so these scenarios will only be navigated successfully if we're paying close attention.

Cosmic Companionship

The future of humanity is uncertain. However it turns out, there are implications for the likelihood of kinship among the stars. All we have to do is assume that the events that led to an intelligent species with technology on this planet weren't a complete fluke.

We've seen that the ingredients for life are universal. The 13.7-billion-year-old universe has also generated many sites for life in the rocky cinders that are left on the cool periphery when a star forms. A best current estimate is about 100 million habitable planets in the Milky Way, which is just 1 among 50 billion galaxies. That's a lot of real estate for potential biology. The question is, how many intelligent, communicable civilizations are there in the Galaxy? In other words, how many alien pen pals might we have?

Astronomer Frank Drake formulated the central equation in the search for extraterrestrial intelligence back in 1961. He recognizes that it's more of a framework for our ignorance than a tool where we can plug

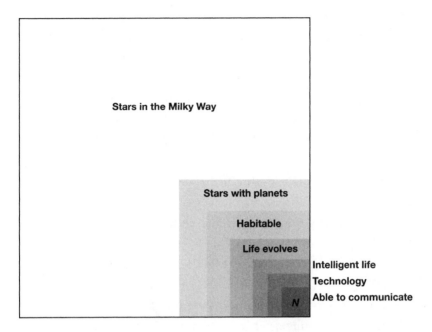

Figure 4.5. The Drake equation starts with the number of Sun-like stars in the Milky Way galaxy, then winnows down the number by successive factors representing stars with planets, planets that have life, life-forms that are intelligent, and intelligent life-forms that can communicate across space.

in hard numbers to get a specific and accurate answer. To begin with, we don't yet know where those millions of habitable planets are. In fact, the roughly 400 planets we know of beyond the Solar System are mostly uninhabitable gas giants. However, computer simulations make astronomers confident that the terrestrial planets are out there waiting to be discovered; we're just getting the first tools powerful enough to find them. Further, we must assume that "the others" will use their technology to communicate or signal through space and not just to improve their lives on the home planet.

That said, to solve for N in the Drake equation, were N represents other civilizations, we take the product of: the number of Sun-like stars born per year, the fraction of those stars that have planets, the number of Earth-like or habitable planets per system, the fraction of those planets that actually harbor life, the fraction of those that continue to develop intelligent life, the fraction of those intelligent species that learn how to communicate through space, and the average longevity of species that have technological, communicating civilizations (Figure 4.5).

That's a whole heap of uncertainty.[9] The first few factors in the Drake equation are far better determined than the last few. Assuming five new Sun-like stars per year, 50 percent of which have planets, with two Earth-like planets per system, all of which develop life but only 20 percent of which evolve to intelligence and technology, simplifies the equation to $N = L$. This means that the total number of galactic pen pals is governed by the average longevity of their civilizations. Frank Drake was suitably impressed by the elegant form; his license plate reads "N EQLS L."

The Milky Way has existed for over 12 billion years. Stars are being born and dying all the time. In some locations, biology has only just started on a newly minted planet, while in others life may have been evolving for many millions of years, and in yet others the life has been extinguished by the death of the parent star. In many more situations the planet exists but never hosts life of any kind. The reduced form of Drake equation distills these myriad possibilities into a simple equality. It's a mirror we hold up to see our reflections in the stars.

What matters isn't the time a species is intelligent but the time it has with technology good enough to send signals into space or listen to signals from analogous civilizations. Notice that the Drake equation focuses our attention exclusively on species that make the jump from self-awareness to technology. That's overly prescriptive. We share Earth with species that have high cognitive function and self-reflection without being able to bend nature to their will—elephants, octopuses, and orcas come to mind. In anticipating the wondrous possibilities of biological evolution out in space, are we really more interested in the use of power tools than the creation of poetry?

If intelligence is enduring but technology is a transient phase in a creature's evolution, then the Drake equation will yield a pessimistic estimate of companionship, perhaps by orders of magnitude. For now let's ignore the species who might be trapped with their own thoughts.

Consider two possibilities. We've clearly come close to midnight on the Doomsday Clock. Suppose that we get to Kurzweil's singularity and the instability of such technology destroys us. If that happens in 50 years it will have been 100 years since we started sending radio signals and spacecraft into space, so $L = 100$ years and $N = 100$. With a hundred civilizations sprinkled thinly through space, the Galaxy is a very lonely place. The average distance between pen pals will be around 10,000 light-years. Worse, they only live 100 years in a communicable state so no signals can be exchanged before one or other of the civilizations self-destructs.

A more optimistic calculation assumes that we'll endure for the typical duration of a species in our currently advanced state. That gives us a million years so $L = 1,000,000$ and $N = 1,000,000$. In this case, the Galaxy is teeming with civilizations and the average distance between them is just 100 light-years, so communication would be an easy task for a patient civilization. We've had blinders on in only paying attention to the few hundred billion stars in the Milky Way. The number of smart aliens in our Galaxy must be multiplied by the 50 billion galaxies in the universe. Most of them are at vast distances, but if some civilizations last tens or hundreds of millions of years, then their conversations can span galaxies.[10]

If they exist, such species are so far beyond the singularity they would seem immortal to us, like gods. Cosmic companionship, or the promise of salvation by a superior entity, probably isn't enough motivation for us to get through our troubled technological childhood, but if we do we can take our place among the progeny of the stars.

Chapter 5

The Man Who Said the Earth Lives inhabits a bucolic setting in rural England. Down a small country lane in the heart of Devon is an old mill and beside it, a slate-roofed cottage. James Lovelock lives here with his second wife and his youngest son, who is mildly disabled. They are surrounded by 30 wooded acres. It's Earth in its most natural state.

When Lovelock was young, the countryside was a sanctuary. He grew up poor and working class in south London. His father had spent six months in prison as a youth for poaching a rabbit to supplement a meager diet. His mother was an early feminist. He was raised by his grandparents. In school he was an inattentive student, dyslexic, more interested in pranks than homework, disappearing for days on end in the imagined worlds of Jules Verne and H. G. Wells. But when his father took him on long country walks to escape the soot and grime of London, the experience transformed him. It's when he says he first saw the face of Gaia.

Lovelock was one of the people who jump-started the environmental movement. Working at his kitchen table in 1957, he invented a small device that could measure minute concentrations of chemicals and pollutants in the air. Five years later, Rachel Carson raised the alarm on pesticides in her path-breaking book Silent Spring. Soon, Lovelock's device was used to attach hard data to Carson's claims, and show how, for example, a factory in Japan could affect the air quality in Europe.

The idea of Gaia came to Lovelock on the heels of a midlife crisis. He was in his midforties, working in a job that paid the bills but left him bored. He had four children at home, one with a birth defect, and a cranky, aging mother. He smoked, he drank, he was unhappy. So when a letter arrived out of the blue inviting him to the Jet Propulsion Lab (JPL) in California to work on planetary probes, he leaped at the chance.

The scientists at JPL were designing instruments for the twin Viking missions that were due to land on Mars and test for life. Their strategy was to dig up the soil and test for bacteria. But, Lovelock thought, why not test the atmosphere instead? If life were there, it would use raw materials in the atmosphere and deposit its wastes back there, just as life on Earth does. It should be easy to detect the imbalance caused by life. A shifted perspective caused him to think about his own planet in a new way. Seen from afar, Earth's surface is painted with life but that life exists in a balance with the atmosphere. His insight was to ask questions about the atmosphere that hadn't previously been asked.

We all take our first breath of life-sustaining atmosphere and then take it for granted the rest of our lives. We're as confident that it will be un-

changing as we are of the rising and setting of the Sun. It's a perfect stained-glass window onto the world but it's also a mixture of unusual, combustible gases. Lovelock's flash of enlightenment in 1965 was the thought that the air we breathe keeps a constant composition and so something must be regulating it. That something is life.

The Restless Earth

Planet with a Fever

Suppose an alien civilization is patrolling our sector of the Galaxy with sentient probes designed to detect life in its myriad forms. Thousands of tiny robotic sentinels swarm through the Solar System, landing on every plausible target and reporting back their analysis at light speed. Imagine exactly a hundred of the probes land on random positions on the Earth like a light, metallic rain. What would they find?

About 70 of them would land on oceans. The instruments would find a billion cells in a typical liter of seawater, comprising over 20,000 bacterial species, with more DNA in a teaspoon of seawater than in a human genome. Roughly nine would land in grasslands and sample from hundreds of plant and animal species. Seven would land in deserts, which have few animals but dozens of plants and hundreds of bacterial species. The six that landed in rainforest would be treated to a feast of biodiversity: 50 percent of the planet's species and hundreds of different plants and animals in plain view. About four would land in the tundra biome, with few animals but many plants, lichens, mosses, and bacteria. Roughly three would land on ice or in glaciated regions. Even there, the frozen landscape would yield traces of biology; bacteria are active in the Antarctic ice at $-20°C$ ($-4°F$).

The probes would succeed whether or not they saw us. In the time of Shakespeare, London was the only city of a million people; now there are over 400 and half the population of the world lives in urban areas with over 40 people per square kilometer. But they cover less than 5 percent of the land surface so on average only 1 of the 100 probes might see an obvious imprint of intelligent life. It doesn't matter. The other 99 would detect creatures large, small, and microscopic, because life grips the planet like a fever.[1]

Earth's early history supports the idea that life is both persistent and durable. Volcanic activity and plate tectonics have erased most of the evidence of the first 700 million years after Earth formed, called the Hadean period. Heat and pressure alter rocks and also eradicate the subtle evidence of microbial life, so there's very little to go on in the early phase. If you pick up a rock wherever you happen to live, odds are it's at most 1 or 2 hundred million years old. Geologists go to a few special places to search for primeval rocks and then look for ancient life inside them.

The oldest rock samples where there's no controversy over the age come from an igneous formation near the Acasta River in Canada's Northwest Territories, where the outcropping is 4.0 billion years old. Recently, rocks as old as 4.3 billion years have been found nearby in northern Canada. The oldest rocks with uncontroversial evidence for life are from Western Australia: fossil microbe "burrows" 3.35 billion years old and fossil microbial colonies called stromatolites 3.5 billion years old.

Before that, events are shrouded in uncertainty over the difficulty of interpreting the limited physical evidence. The conventional wisdom has been that the Hadean matched the Christian view of Hell, a place of fire and brimstone. The conventional wisdom also held that water was mostly delivered by a late epoch of bombardment by asteroids and meteors, and that life couldn't have started until that chaos was over, about 3.8 billion years ago. But recent research suggests that early Earth matched the Hades of ancient Greek myth: a cool, misty, and gloomy place. Here's what we think we know.

Earth formed 4.55 billion years ago and the early action was intense. Within 30 million years—0.5 percent of the infant planet's life—it was fully assembled as dust grains grew to boulders and then mountain-sized rocks, those large pieces assembled under gravity into a huge molten world, distinct chemical layering of the mantle took place, and a Mars-sized object slammed into the infant planet to create the Moon. Whew.

Remarkable insights have come from the oldest materials found on Earth—not rocks but tiny zircon crystals from Western Australia. The oldest specimen is 4.4 billion years old and many are more than 4.2 billion years old. Minerals within the crystals point to a watery world with a solid crust and normal rock recycling processes.[2] Because the Sun was 30 percent dimmer then, Earth was cool and parts of it may have even been covered with ice. With this new picture, it's plausible that life began within a mere hundred million years of formation, a billion years before the earliest evidence in hand.

It wasn't smooth sailing after that; in the next half billion years the planet suffered through a series of impacts as the outer planets took their current positions and sent a wave of asteroids on Earth-crossing orbits. Calculations by geophysicist Norm Sleep at Stanford suggest a dozen objects 150 kilometers across slammed into us. Three or four were 300 kilometers across, with collisions violent enough to boil off most of the oceans. After such huge impacts, the atmosphere would have filled with rock vapor and steam, and it would have taken the oceans thousands of years to return to their normal levels (Figure 5.1). Geologists don't agree on what constitutes a "sterilizing" impact, but even life in the deepest parts of the ocean or inside the crust might not have been able to survive the mayhem of the largest impacts.

If Earth has had a biosphere for at least 95 percent of its history, spreading its tentacles to every nook and cranny, then biology appears to be a natural condition for the planet. Which leads to an important question: how did it get started?

Something from Not Much

The workings of the simplest cell seem so complex and profound, it's no wonder people without biology training swoon into the creationist and intelligent design schools. It's not quite ex nihilo but the biosphere was

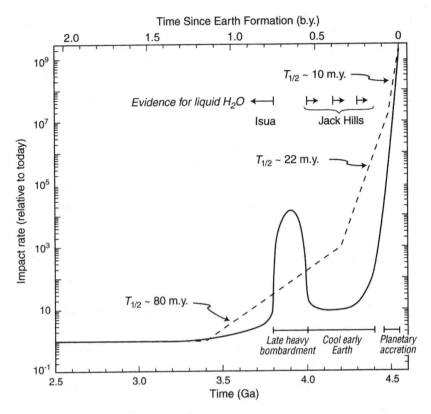

Figure 5.1. Impact rate over the first 2 billion years (b.y.) of Earth's history. The dashed curve shows exponential decline in three stages, with a time for the rate to fall by 50 percent ($T_{1/2}$) of 10, 22, and 80 million years (m.y.). The solid curve shows the favored model of a very rapid decline followed by a late bombardment epoch. The evidence for water in the zircons from Jack Hills, Western Australia, argues for a hospitable and cool Earth within 200 million years of its formation. The bottom of the graph is labeled in billions of years ago (Ga).

created from simple forms of inanimate matter, which is not very much. The question, how did life begin? begs the question, what is life? and a deeper question lurking right behind it: why life?

For over 2000 years people believed that life arose spontaneously and routinely from inanimate matter. Aristotle thought it an unremarkable truth that maggots arose from rotting meat, mice from hay, birds from trees, lice from sweat, and crocodiles from rotting logs. This view was discredited by simple experiments in the mid-seventeenth century, and it was finally refuted when Pasteur provided support for the germ theory and the idea that living things grow from other living things.

In the middle of the nineteenth century, Pasteur explained the reproduction and evolution of microorganisms and Darwin explained the
analogous reproduction and evolution of macroorganisms, which still
begged the question of how inanimate matter turned into life in the
first place.[3]

Research on the origin of life is historical science and not experimental science. There are few direct traces of the first billion years of
Earth's evolution, and it's hard to tease subtle biological signatures from
rocks that have already been tortured to within an inch of their lives.
It's surprising we have testable theories at all, given the amazing events
that turned mineral into muscle, clay into camouflage, bacteria into
birdsong, and modest molecules into DNA's soaring spiral staircase.

Darwin skirted the issue in *The Origin of Species*, but in a letter
to a colleague in 1871 he suggested that life may have started "in a
warm little pond, with all sorts of ammonia and phosphoric salts, lights,
heat, electricity, etc. present, so that a protein compound was chemically formed ready to undergo still more complex changes." In the
1920s Russian biochemist Alexander Oparin and British evolutionary
biologist John Haldane developed the "primordial soup" idea, where
simple organic molecules could become more complex in the primordial
oceans, without oxygen but with the assistance of sunlight. American
chemist Harold Uery and his student Stanley Miller tested the Oparin-
Haldane hypothesis with a classic "life in a bottle" experiment in 1952,
where the chemical ingredients and the energy conditions of the early
Earth reacted in a closed flask. No life was produced but many of the
building blocks resulted. In an interesting recent twist, one of their
unpublished experiments was reanalyzed and the flask, where the conditions had been designed to simulate a volcanic eruption, surprisingly
contained 22 amino acids.

When did life start? Surface life may have had its first origin as early
as 4.2 or even 4.4 billion years ago, only to be obliterated by impacts.
The "impact frustration" of biology may have happened multiple times.
Rocks from the Issua formation in Greenland show an uptake rate of
radioactive carbon that's suggestive of a metabolism at work 3.85 billion
years old, just after the end of the heavy bombardment. The evidence

isn't unequivocal. Bacterial colonies are seen in the fossil record just 300 million years later. Life that emerged near deep-sea volcanic vents could have survived impacts, but there's a persistence problem. Hydrothermal vents aren't long-term stable environments; they come and go and life can't easily spread by traveling from one to another.[4]

Time may not be a problem because the transition from primordial soup to simple organisms could have occurred rapidly. One study by Mexican biologist Antonio Lazcano and Stanley Miller suggested that self-replicating systems capable of Darwinian evolution can emerge in less than a million years,[5] and simple cyanobacteria may have developed with a gene duplication rate of one per thousand years. That would be less than 10 million years to go from soup to bugs.

There's no single accepted theory for how life began. Some groups, like that led by Craig Venter at the Institute for Genomics Research, have tried a top-down approach, engineering prokaryotic cells with fewer and fewer genes until the minimum requirements for life are met.[6] Other groups, such as the one led by Jack Szostak at Harvard University, are using a bottom-up approach, trying to synthesize a protocell from simpler components.

In going from nonlife to life, the hardest step seems to be getting simple organic building blocks to turn into polymers and then on to complex structures that interact consistently to form a primitive cell. As an example of problems encountered in this research, water is a useful solvent, dissolving and transporting chemicals that carry out many biological functions. But water also breaks down polymers as opposed to building them up, which acts against greater complexity. Another perennial issue is that modern cells require the cooperative action of proteins and nucleic acids, and can't function without either. So how do we know which came first?

Many biologists think that the versatile information-carrying molecule RNA was a precursor to DNA. As a result, research has been done on what's called the "RNA world" hypothesis, a hypothetical time before the first cells when short strands of RNA (or an even more primitive cousin) catalyzed their own replication. Growth can take place on clay or pyrite surfaces or within spontaneously forming fatty microspheres.

Darwinian selection has been shown to operate on the self-catalyzing structures, so they would outcompete less useful molecules. Think of Lego machines that can add to themselves, and then the results play in an arena where the most effective or robust machines survive.

Only time will tell whether RNA world is a "just so" story or a compelling scenario for how life began. Meanwhile other novel ideas are in play. Günther Wächtershäuser's "iron-sulfur world" hypothesis says that metabolism predates genetics. In his model, the key idea is that biochemical complexity builds up not in the open oceans but in hydrothermal vents. The recipe for life? Boil water. Stir in iron sulfide. Bubble in carbon monoxide and hydrogen sulfide. Wait for amino acids and proteins to form.

More recently his model has been given impetus by researchers who claim that the first cells were not living but were inorganic, made of iron sulfide formed in total darkness on the ocean floor. Microscale cavities near hydrothermal vents can act as containers for chemical reactions, concentrating ingredients yet with a constant flow through of hydrothermal fluid. In addition, Carnegie Institution geologist Robert Hazen and UC Santa Cruz chemist David Deamer have found that when a solution of pyruvate in water is subjected to conditions like those near a hydrothermal vent, cell-like vesicles can form spontaneously (Figure 5.2). Pryuvate is a critical compound in biochemistry, at the intersection of a network of metabolic pathways. Perhaps volcanic vent creatures were our "last common ancestors," and this membrane allowed these protocells to leave their caves.

Life on the Edge

Imagine jumping into a pleasantly cool swimming pool on a warm summer's day. Next, imagine the pool is filled with water near the boiling point or just above freezing. Now imagine the pool has been filled with vinegar or household ammonia or drain cleaner or brine or battery acid. In all cases but the first, you'd be in trouble in seconds and dead not long after. But there are microorganisms that not only tolerate each of

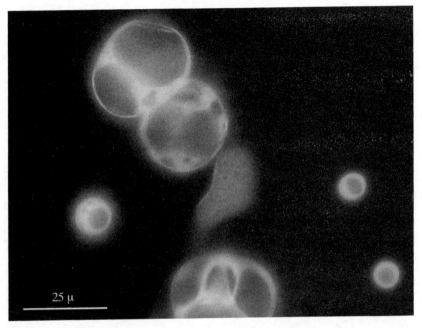

Figure 5.2. When a solution of the biochemically critical compound pyruvate in water is subjected to the high temperature and pressure typical of a hydrothermal event, membranes form and spontaneously organize into cell-like containers.

these conditions, but they also thrive. These tiny creatures are called extremophiles and they've challenged our assumptions about where life can and can't live on this planet.

Chris McKay is friendly and sociable but each year he ventures as far as possible away from any people to search for life in the most unlikely places. His favorite hideouts are Siberia, the Atacama Desert in Chile, and the high Antarctic valleys. This rugged astrobiologist perversely looks for life where it's cold and arid because he wants to find analogs of Mars on Earth. He wants to know what the limits of life on Earth tell us about the best strategies for finding life on the red planet.

Each year, McKay's team flies into the coastal city of Antofagasta in the north of Chile and drives several hours to a high plain that lies in a double rain shadow—it's sheltered by coastal mountains to the west and by the volcanic peaks of the Andes to the east. Passing through the abandoned village of Yungay, the landscape looks utterly barren.

Jagged, bare rocks and salt flats bake under an empty sky. There are places where it hasn't rained in recorded history. If you look closer, however, there's life: patches of lichen, splashes of colorful desert varnish, and even small bugs scurrying in the dirt. But with great perseverance, McKay has found several truly barren spots. "In the driest part of the Atacama, we found that, if *Viking* had landed there instead of on Mars and had done exactly the same experiments, we would also have been shut out," he says.[7]

Meanwhile, Diana Northup and her team prefer to look for life where it's cramped and dark, and they're quite happy to deal with slime and toxic gases. She's a microbiologist and a caver and she finds abundant life 300 meters below the surface, in total darkness. Like the floor of the ocean, caves aren't completely isolated; some organic material and water trickle in. Many of the microbes she finds are Archaea, the branch closest to the root of the tree of life. That makes them modern analogs of Earth's first forms of life.

In Mexico, Northup and her colleagues enter a strange world where life lives on hydrogen sulfide and other noxious chemicals. "Some of the ones I see have long stalks; they look like sperm on testosterone," she says. "Some of them look like braided ropes. They're *really* cool. And the stuff they produce is just incredible. I can go on and on about blue goo and slime balls and "snottites," the slimy bacterial stalactites found in Cueva de Villa Luz." There's danger too. Talking about Cueva de Villa Luz, her favorite spot, she recalls, "There is carbon dioxide at high levels, there's formaldehyde, there's sulfur dioxide, there are lots of other things that our respirators don't protect us against. And, of course, you can die from hydrogen sulfide in a matter of seconds."[8]

Extremophiles are found at temperatures above the boiling point and below the freezing point of water, in water so saline salt precipitates out of it, and in pH conditions ranging from strongly acidic to strongly basic. These microbes can live on top of the highest mountain or in a deep ocean trench at a pressure of 1000 atmospheres. Some can live inside rock and others can go into a freeze-dried wait state for tens of thousands of years. There's even a type of bacteria that can tolerate 1000 times the radiation dose that would quickly kill you or me. Some microbes harvest their energy from methane, sulfur, iron, cadmium, and even arsenic. They do this by evolving biochemical strategies to protect reproduction and information storage in the cell.

Such extraordinary organisms tell us that the biosphere is pervasive and very robust. By finding that life "fills" the environmental envelope of Earth, we can infer that if those physical bounds were made even more extreme, life would adapt to the new range. It's a small step to speculate that life formed and currently thrives in a host of varied and similarly extreme environments throughout the cosmos.

The Shadow Biosphere

Microbes outnumber plants and animals not only numerically but also in terms of number of species. The numbers seriously underestimate how much they dominate the biosphere because so few can be grown in "captivity." Only 1 percent of the microbes in any soil or water sample can be cultured with laboratory techniques because of their intimate connections to their unique ecosystems. It's as if we snatched a Bantu tribesman or an Inuit fisherman and set him down in Manhattan and said, "Go about your business!"

The veil of ignorance over the true nature of the microbial world is being lifted by technology. Stephen Quake is a researcher at Howard Hughes Medical Institute who has gotten around the clumsiness of the standard lab setup by shrinking an entire automated genomics lab to the size of a postage stamp.[9] His microfluidics techniques isolate the organism in a nanoliter of liquid and in this tiny volume he amplifies and analyzes the genetic material of a single cell, meaning he doesn't need to use a traditional lab culture. In just two years, it's been used to study many of the 700 species of microbes that live in our mouths. Another tiny microarray, which can test for 9000 variations of a single gene that appears in all bacteria, has been used to show that there are over 1800 types of bacteria wafting on the air. The biosphere extends upward as well as downward.

We know life is liberally painted on the surface of the crust, extending to places where the crust is under water, but what about deeper? It's been known for decades that extremophiles called endoliths can exist inside rock or between mineral crystals. They use chemical sources of energy so don't need sunlight. We know very little about life under the crust because digging holes is so hard—it took the Russians 15 years to get just 24 kilometers down at the Kola borehole.

Desulforudis audaxviator is a sign of what we might find (Figure 5.3). A species of bacteria discovered 2.73 kilometers down a gold mine in South Africa, it is named after a message decoded by the hero from a well-known book by Jules Verne: "descend, bold traveler, and attain the center of the Earth." *D. audaxivator* gets energy from the decay of uranium and it survives in total darkness, without oxygen, and at a temperature of 60°C (140°F). It's the only organism ever found to live in total isolation; it has everything it needs to eat, move, protect itself from viruses, and sur-

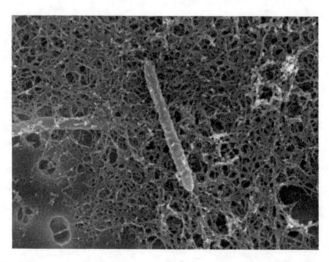

Figure 5.3. The extraordinary rod-shaped bacterium *Desulforudis audaxviator* lives in total darkness miles underground, gets its energy from the decay products of uranium, and makes organic material from the rocks around it. It co-opted genes from ancient Archaea to protect itself from viruses.

vive nutrient-poor conditions, all in a tidy 2200-gene package. It's an eco-system of one, oblivious to the hubbub of the surface world.

In 2002, geobiologist John Parkes captured mud core samples from a similar depth below the seafloor off the coast of Newfoundland. There were simple prokaryotes in every sample. Half of them were alive but their cell division rate was incredibly slow, perhaps once a century. In an environment with few predators and little food, life slows down. He speculated that the microbes might be millions of years old. They may even represent an unbroken line to the beginning of life, because this is a place that could survive impacts by meteors and asteroids.

Another project called the Ocean Drilling Program used deep sea core samples from around the world to infer an amazing 81 billion metric tons of microbial organisms living in the deep biosphere. Bacteria are found near the surface but about 90 percent of this hidden biosphere comprises Archaea, primitive descendants of the earliest forms of life. In this environment they're "starvelings," subsisting on the fossil remains of plants that were predigested by generations of different organisms. The interior biome rivals the conventional surface biosphere in mass.

Carol Cleland isn't worried about microbes that are hard to dig up or sequence; she's worried about forms of life that are so strange we wouldn't recognize them. A philosopher at the University of Colorado, she coined the phrase "shadow biosphere" in 2005 and has written a book on the problem of defining and identifying life. The concern is "a microbial biosphere that's so chemically and molecularly different from life as we know it that it wouldn't be in direct competition with familiar life; familiar life couldn't metabolize it, and it would occupy ecological niches that were under-populated by familiar organisms."[10]

The shadow biosphere is a provocative hypothesis but Cleland is quite serious when she speculates that we could share the planet with alien life-forms. Our methods are designed to detect life as we know it, not life as we *don't* know it. For example, biological sensing equipment on Earth (and eventually to be used on Mars) will use a polymerase chain reaction, the standard method for amplifying small fragments of DNA. An organism that doesn't use nucleic acids for storing and transmitting information will not register as alive with this equipment. Biology not based on carbon or life that uses exotic energy sources might also be missed by standard testing equipment. We look for what we know.

Gaia

Cycles of Life and Rock

To many people, a rock is a rock is a rock, but any visitor to a mineral museum has seen the breathtaking variety and colorful beauty of the terrestrial environment. Research published in 2008 shows that this diversity is intimately connected with the presence and evolution of life. There are just a dozen primordial minerals in the interstellar dust grains that gathered together by gravity to form Earth and the other inner planets. Variations in temperature and pressure are required to make additional mineral species. Small Solar System bodies like moons and asteroids have about 60 different minerals. Planets with volcanoes and some water like Venus and Mars are estimated to have 500 types of mineral in their surface rocks. Earth has had the extra ingredient of plate tectonics where heat and stirring in the interior "kitchen" cooked a total of nearly 1400 mineral types.[11]

However, only Earth among the Solar System bodies hosts life, and its biological processes tripled the number of minerals around us to 4300. Microscopic algae produced oxygen in the atmosphere and weathering of the oxidized rocks produced ores of iron, copper, and many other metals. Microbes and plants also facilitated the generation of diverse clay minerals, and mineralized skeletons and shells made minerals like calcite, which would be rare on a lifeless planet. For nearly 3 billion years, geology evolved in parallel with biology. The effects of life on rocks are so profound and pervasive that astronomers hope to use remote sensing of minerals on other moons and planets to identify biology beyond Earth.

There's a grainy connection between rocks and life, like the warp and weft of a fabric. But geology and biology are also connected on the large scale, at the level of whole garments. We can make sense of it by following the interactions and transactions of the three key players: oxygen, carbon, and nitrogen. They're the third, fourth, and sixth most abundant elements in the universe, after the über gas hydrogen and the noble gas helium (neon, the flashy interloper, is number five). Carbon is the "grit" of life, the universal Lego block for building complexity. Oxygen is the reactive and volatile player, clinging to other elements to make rocks, rusting metals and exerting oxidation stress on many creatures. Nitrogen is oxygen's aloof cousin, essential for biology but unwilling to be part of the game unless bribed with a lot of energy.

Carbon is part of two cycles, a geological one that spans millions of years and a biological one that can last as much as a thousand years and as little as a few days. Carbon dioxide dissolves in seawater to make a weak acid that combines with calcium and magnesium to form carbonates, a process called weathering. Erosion washes carbonates into the ocean and they settle to the seafloor. The material is drawn into the mantle by subduction as one rocky plate slides under another. The cycle is completed when carbon dioxide is liberated in the heat of the mantle and belched into the atmosphere by volcanoes. There's a huge carbon reservoir in the crust and the oceans, giving a long-term stability to the CO_2 content of the atmosphere. It's like a gambler with a lot of carbon "chips" who only plays a few at a time (also see Figure 2.6).

Living things have a hundred times the concentration of carbon as Earth's crust. Life extracts carbon from its environment as plants and algae use light energy to convert carbon dioxide into carbohydrates. Plants and animals extract energy from the carbohydrate "fuel" and respirate the carbon dioxide back into the atmosphere. Additional CO_2 is returned by burning and decay. The carbon involved in the biological cycle each year is a thousand times more than the amount involved in the geological cycle. This is like a gambler who bets a sizable portion of his chips at any time.

There's also long-term storage of carbon via biology. Land vegetation extracts carbon dioxide from the air when soil weathers. During times when photosynthesis exceeded respiration, organic material built up slowly to make oil and gas deposits. And in the oceans, some of the carbon taken up by plankton passes through the food chain into sea-shells which settle to the ocean floor to form sediments.

Our current problem with global warming stems from the fact that we're rapidly reversing the first two processes by burning and cutting down forests and we're voraciously consuming our stockpiles of fossil fuel. The cycle isn't in balance because the return of CO_2 exceeds the uptake. It's almost certain that human activity has caused the carbon dioxide level to increase 20 percent in the past century, cranking it higher than at any time for half a million years. That's 10,000 times faster than a natural rate of increase and Earth's carbon thermostat can't keep up. Continuing the gambling analogy, we're drawing from our savings to play, losing quite a lot, and getting hot and bothered about it.

The Primal Goddess

James Lovelock is a shy maverick. He never intended for his idea to have mythic status and cross over into the popular culture. But when he lived in a small Wiltshire village, one of his neighbors was the novelist William Golding, noted for *Lord of the Flies*. On a walk one day to the post office, Golding suggested the name Gaia for the theory. It stuck. Hesiod's poetry from the seventh century BC names Gaia as a primor-dial god, the first one born out of Chaos, the original void of space. Gaia personifies Earth. In ancient Greece, oaths sworn in the name of Gaia were considered the most binding of all.

Gaia is best summarized in Lovelock's own words, here from an article in *Nature* in 2003: "organisms and their material environment evolve as a single coupled system, from which emerges the sustained self-regulation of climate and chemistry at a habitable state for what-ever is the current biota." Lovelock was struck by the constant tem-perature of the Earth over billions of years despite an increase in the energy of the Sun by 25 percent. He also noted the composition of the atmosphere didn't change, despite the presence of a large amount of oxygen, an unstable gas that should quickly combine with minerals in the crust. Finally, he was puzzled by the constant salinity of the oceans, not moving beyond the narrow range required for cells to function, despite a level of river runoff that should raise it to much higher levels. Lovelock speculated that there was a global control system causing this stability.

The complexity of climate is caused in part by feedback processes. In negative feedback, a signal fed back into a system is reduced or canceled so that change is minimized. A thermostat is an example. In positive feedback, a signal fed back into a system is amplified so change increases rapidly and maybe uncontrollably. A nuclear chain reaction or the ear-splitting feedback in an amp are examples.

The nexus of earth, water, and air causes a number of feedback loops. If surface temperature rises, more water vapor goes into the air, which increases low cloud cover and reflects more sunlight, causing cooling. On the other hand, if the temperature falls, less water vapor goes into the air, decreasing cloud cover and causing heating. That's negative feedback. But it's not that simple because water vapor traps heat, so more water vapor in the air raises the temperature, which evaporates more water, which raises the temperature even more. That's positive feedback (Figure 5.4). Which one wins? It all depends on cloud cover, which is one of the hardest things for scientists to realistically include in computer models.

Lovelock's hypothesis, confirmed in part but not in detail by data and models, is that life plays an intimate role in these cycles, in the sense of negative feedback, or reducing the variations. Suppose Earth goes through a geologically active phase where volcanoes belch out lots of CO_2. It's removed from the air by weathering when it interacts with minerals in rocks, a process that's accelerated by soil life. When the CO_2 is dissolved in seawater, some of it is trapped in the creature's shells, which get entombed on the seafloor when they die. CO_2 is also consumed by algae in the top layer of the ocean, and they release a gas when they die that nucleates water drops and increases cloud cover. That makes at least four different ways that life moderates an increase in CO_2 from volcanic activity.

Gaia gained extra impetus when Lovelock began working with Lynn Margulis, the microbiologist who came up with the widely accepted

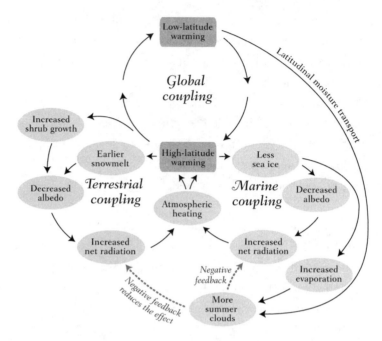

Figure 5.4. Global climate is governed by a complex set of feedback cycles, some of which damp or reduce changes (negative feedback) and some of which amplify or increase changes (positive feedback). The overall effects are difficult to predict because of the complexity and because inputs like vegetation, cloud cover, and sea ice aren't easy to measure or model.

theory that the eukaryotic cell is a symbiotic union of primitive prokaryotic cells. Margulis views Gaia as symbiosis as seen from space. Life exists in a symbiotic relationship with the environment and, as a result, the environment is self-regulating so as to sustain living organisms. Gaia acquired New Age trappings, and pushback from many biologists and ecologists, because Lovelock referred to Earth as an organism. It has never been clear how natural selection could operate at the level of a planet so this "strong" form of Gaia isn't widely accepted.

And what now, when positive feedback may be acting so strongly to increase temperatures that the natural thermostats are broken? Lynn Margulis is an optimist. "Gaia is a tough bitch," she says. Lovelock, by contrast, is a cheerful pessimist. "Enjoy life while you can," he says, "because if you're lucky it's going to be twenty years before it hits the fan." One by one, he swats down the most popular ideas for ecological living. Carbon offsetting? It's just a joke. Paying money to plant more trees? You're probably making it worse. Recycling? Almost certainly a waste of time and energy. Green lifestyle? Just an ostentatious grand gesture. Lovelock thinks global warming is irreversible and nothing can stop large parts of the planet from becoming too hot to inhabit, sinking underwater, or causing mass migration, famine, and epidemics.[12]

Mother's Cold Shoulder

In the face of such gloom, it's worth remembering that life survived a lot worse than we'll experience in even the worst warming scenarios. The current global warming is severe because it's a result of several kinds of positive feedback. There's melting ice which decreases the amount of reflected sunlight. There's more water vapor from the oceans which increases the amount of trapped heat. There's melting tundra causing the release of methane, which is a potent greenhouse gas. And there's desertification which reduces the amount of vegetation sequestering carbon dioxide.

But what goes up can go down. Picture this: White vistas stretching to the horizon on what once had been ocean. An ice pack a thousand feet thick and the temperature a numbing $-32°C$ ($-25°F$). Sheets of ice that almost meet at the equator. Continents where water and ice have evaporated, rain has stopped, and weathering has slowed to a crawl. The landscape is stark, nothing but sterile brown rock. These would have been scenes of Earth to a time traveler 700 million years ago.

It's called Snowball Earth, and while scientists aren't exactly sure what triggered it, there are ways that negative feedback could have driven the planet into the deep freeze. In this time before the separation of the continents, the large equatorial land mass would have absorbed more sunlight, with a high rate of weathering removing lots of carbon dioxide from the atmosphere. As it cooled and the ice pack grew, the increased ice cover would have reflected more sunlight, increasing the cooling. Evidence for Snowball Earth comes from glacial deposits seen at equatorial latitudes and a set of sedimentary formations consistent with ice-covered oceans.[13] Life was severely stressed during this chilly spell because the land surface was dry and barren and ocean life was mostly deprived of photosynthesis.

How did we escape? Even though the normal hydrological cycle barely operated, the planet's volcanic engine was working normally. Carbon dioxide pumped out by volcanoes increased the temperature just enough for the ice pack to develop fissures and cracks. This released methane—a potent greenhouse gas—produced by microbial life in the oceans. As the ice eroded, weathering dumped nutrients like phosphorus into the oceans, causing a surge in cyanobacteria. The new dark areas of water absorbed more sunlight, and heating accelerated. With all this positive feedback, Earth would have emerged from the icy grip in as little as a thousand years.

Early speculation that Earth was completely covered by ice, like a frozen billiard ball, has given way to a more complex picture, with ice ages and waves of glaciation from 790 to 630 million years ago, where the three most severe episodes cloaked most of the planet in ice. There is some evidence for a much earlier Snowball Earth episode, 2.2 to 2.3 billion years ago, during the Paleoproterozoic era when the continents were first forming. It's striking that the two snowball phases, when biology was severely stressed, preceded a surge in the atmosphere's oxygen level and then in the complexity of life in the oceans.

Biomarkers

Volcanism and plate tectonics are the forces that keep a planet active and, as we've seen, they contribute to mineral diversity. Small moons and planets don't have enough mass to retain an atmosphere or keep a molten core that can drive a changing crust, so astronomers suspect they're biologically as well as geologically dead. We come full circle to James Lovelock's insight when he was working on the Viking mission. He concluded that it was unnecessary to send a spacecraft to Mars. All you had to do was find out if the Martian atmosphere was in chemical equilibrium. It was, and so Lovelock concluded that Mars is dead. The dynamism that points to coupled biology and geology is missing.

A useful way to understand the durability of biospheres is to imagine Earth as seen from afar. What are the telltale signs that this is a living planet (beyond the recent intrusion of humans and civilization)? Astronomers call these signatures biomarkers and they're central to determining the best strategies for finding life on the vast number of planets beyond the Solar System.

When I was 25 and a postdoc at the University of Hawaii, my mentor took me aside. He was an observatory director, highly accomplished, and at the height of his career. Eric was of Viking stock, with a long mane of blonde hair and larger than life in many ways. You'd better be enjoying yourself, he told me. As a grad student you didn't know much and you had a thesis to write. If you get a faculty job, you'll be buried in committee work and have to write grant proposals to fund your research. In a life of science, he said, this is as good as it gets.

Lisa Kaltenegger is having the time of her life. She's a postdoc at the Smithsonian Astrophysical Observatory and a lecturer at the nearby Harvard astronomy department. She goes to meetings and gives talks around the world: Aspen, Santa Clara, Les Houches, Berlin, Frascati, Vancouver, Santiago. A tall and striking Austrian, with a mellifluous middle-European accent, she likes to dance and ride horses, when she can find the time. Mostly she likes to think deeply about how to detect planets in other solar systems, and how we would ever know if they're living planets.

When viewed from a great distance, a planet reduces in size to a dot and the complexities of geology, atmospheric chemistry, and biology must be inferred from the limited information contained in a spectrum. This approach was inspired by Carl Sagan, who wrote a paper in 1993 analyzing a spectrum of Earth taken by the Galileo probe as it headed for the outer planets (Figure 5.5). Sagan thought that the presence of oxygen and methane, plus a red-absorbing pigment with no mineral origin, would be strongly suggestive of biology. Kaltenegger refined this work, tracking the subtle changes to Earth's biomarkers as life evolved.

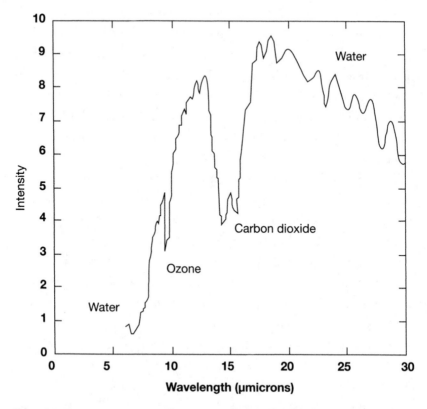

Figure 5.5. The spectrum of a habitable planet. This is Earth as observed by *Mars Global Surveyor*, looking back. Small concentrations of carbon dioxide show clearly, as does water, but the best biomarker is oxygen, seen in this infrared spectrum as strong absorption by the associated ozone in the upper atmosphere.

Her first snapshot is Epoch Zero, 3.9 billion years ago, probably the time when life-formed. The turbulent, steamy atmosphere is made of nitrogen, hydrogen sulfide, and lots of carbon dioxide, which keeps the planet warm despite the dim Sun. Life exists in pockets, but it has to endure heavy bombardment so it leaves no large-scale traces. Epoch One, about 3.5 billion years ago, sees the rise of prokaryotic bacteria which begin changing the atmosphere by consuming carbon dioxide and releasing methane. There is hundreds of times more methane than in our atmosphere today, making it an excellent biomarker for the early phase of an Earth-like planet.[14]

Epoch Two, 2.4 billion years ago, is when cyanobacteria learn the incredible trick of turning water, carbon dioxide, and light into oxygen and sugar. Photosynthesis is very efficient but if it had taken hold too quickly it would have poisoned the microbes creating it. Luckily, there is a small supply of oxygen produced by ultraviolet radiation reacting with glacial ice, which is enough to facilitate the evolution of enzymes to protect against the corrosive gas. And so the oxygen "revolution" is underway. But not without a hiccup; the oxygen consumes methane and the loss of that greenhouse gas plunges us into a Snowball Earth. After the recovery, Epoch Three sees the biological innovations of cells that could breathe oxygen as we do and multicelled organisms.

During Epoch Four, 800 million years ago, oxygen in the atmosphere rises again to reach present-day levels and the Cambrian explosion of life in the oceans occurs after Snowball Earth episodes clear out many new ecological niches. Carbon dioxide levels fall. There are swamps, a few volcanoes, and a single supercontinent rimmed by shallow seas. It is only at Epoch Five, 300 million years ago, that life moves onto land and the atmosphere reaches its current composition. In this last span of time, green vegetation is widespread enough to show in a spectrum as a chlorophyll "edge" but otherwise the key biomarker is oxygen.

Kaltenegger knows that human technological prowess is just a blip on the radar of time in 4 billion years of evolution. "I'm sorry to say the first signs of E.T. probably won't be a TV or radio broadcast," she says. "Instead, it could be oxygen from algae." She also knows that choice real estate like ours is rare among the first 400 known planets beyond the Solar System. "Good planets are hard to find," she warns.

Chapter 6

THREATS TO THE BIOSPHERE

Earth is unfamiliar and exotic. The Four Corners area of New Mexico is a swamp at the fringe of a shallow sea that stretches south across Texas and east to Missouri. The sea teems with life: 10-foot-high penguins, 40-foot-long elasmosaurs, fish, and sharks. The land is tropical and verdant, coated with redwoods, willows, magnolia, and roses. Thunder lizards are at the top of the food chain, with Tyrannosaurs the dominant predator. For the first time in history, birds are becoming common, as are mammals, which remain small and primitive. There are no polar ice caps and North America is 15 degrees warmer than it is today.

By a quirk of fate, Earth's motion in time and space takes it across the trajectory of a city-sized lump of rock moving at 40,000 miles per hour. Catastrophe arrives from a cloudless sky one day 65 million years ago. In nature's roll of the dice, most large reptiles are obliterated by the force of the impact, or by the earthquakes and tidal waves it unleashes. Sea life dies as the atmosphere is filled with Sun-shrouding dust, which destroys the base of the microbial food chain. Small mammals are the primary beneficiaries of the newly evacuated ecological landscape.

In a gradually cooling world where reptiles are scarce, mammals diversify and spread across all the continents. Of 10 mammal families before the impact, 5 are extinguished but those remaining grow to nearly 80 families within 15 million years. The new mammals are every bit as impressive as the reptiles they replace; they include plant eaters 20 feet high at the shoulder and saber-toothed cats as big as modern elephants. By 6 million years ago tree apes are found across three continents. The surge in brain size that led to humans in the past million years has never been seen in any other animal lineage. The hairless ape called Homo sapiens *masters the planet in a few short millennia. It's a species that seems able to indemnify itself from the force of natural selection.*

But what if the passage of a nearby star had altered the gravitational environment of the Solar System enough that the huge rock sailed harmlessly by 65 million years ago? We imagine it this way. The dinosaurs— diverse and dominant for over 150 million years—continue their reign. Mammals remain shrewlike and furtive in a reptilian world. Apes, primates, and humans never evolve. Evolution ebbs and flows uneventfully.

And what if we don't rewrite history, but instead learn of an Earth-crossing asteroid with a projected arrival a few years away? It would be a cruel twist of fate if our budding technology were unable to fend off disaster, allowing us to contemplate the demise of our species without being able to prevent it. When death arrives from the cloudless sky to start a new chapter in history, many species will be lost but only one will feel the loss acutely.

A Hard Rain

Space Rubble

Imagine you're out in space halfway between Earth and the Moon. Looking homeward you'd see the home planet, the Pale Blue Dot. Its delicate beauty set against the blackness of space is an unforgettable sight. You'd also be aware of the utter stillness. The Solar System is almost completely empty.

But looking outward to the Moon you'd notice its pockmarked surface. Craters cover its surface, large and small, so many that there are lots of craters within craters. The Moon is like a mirror. We're so close that its journey through the Solar System and the Galaxy is our journey. If the Moon has been exposed to enough debris to pepper it with craters then that's been our situation too. The Moon presents a perfect record of the past because it has no atmosphere and it's too small for active geology. Erosion, weathering, and plate tectonics give our planet amnesia about its violent history.

Most space junk comes from the asteroid belt, the ring of rocky debris between the orbits of Mars and Jupiter that represents a failed planet. Asteroids can break apart, collide, or interact gravitationally, and any of these processes sends a small fraction on inward trajectories that can cross Earth's orbit. A minority of the space junk that hits Earth consists of comets or cometary debris. Comet impacts tend to be very damaging because they approach at speeds of up to 50 kilometers per second (100,000 miles per hour), compared to speeds of 10 to 20 kilometers per second (20,000 to 40,000 miles per hour) for asteroids.

An aside on terminology. If space rubble is smaller than the diameter of a hair, it's interplanetary dust. If it's bigger than a hair but smaller than a bus, it's a meteoroid. If it's larger than a bus, it's an asteroid (unless it's a comet). The visible path of a meteoroid that enters Earth's atmosphere is called a meteor or a shooting star. Meteoroids that hit the ground are called meteorites. Many meteors are a shower. A fireball is a brighter-than-average meteor as seen by an astronomer, but if it's spotted by a geologist it's called a bolide. The most primitive meteoric material comes from planetesimals. Meteorites that contain tiny spheres of rock are called chrondrites (there are four types—don't ask). Meteorites without those spheres are achondrites. There are also iron meteorites, a group called pallasites, and a couple of dozen from the Moon or Mars. Astronomers are fussy about all these distinctions. There'll be a test later.

This sounds very esoteric and abstract, but interstellar debris is real. About a hundred tons rain down each day, and luckily most bits are very small. You can share in the fun of gathering it; all you need is a bucket, a magnet, and a microscope. Micrometeorites are falling on our houses (and on us) all the time. Put the bucket under a rain spout and wait for space dust to be washed off your roof by rain. Remove all twigs and leaves and spread the rest on a plastic sheet. By passing a strong neodymium magnet over the sheet you'll pull out the magnetic particles. Look at them under a microscope. The micrometeorites are small and rounded, with tiny pits as evidence of their fiery trip through the atmosphere.

Minnows and Whales

The tiny pieces of cosmic dandruff that settle onto our heads and our shoulders are no cause for concern, but their much larger brethren are. The greatest hazard to Earth came immediately after it formed, when planetesimals were still growing to form the modern day planets. Within a hundred million years, most of the material in the cool solar nebula was mopped up. Earth was nearly destroyed by the impact that created the Moon, but then things settled down. After a pulse of impacts during the heavy bombardment the impact rate has fallen to levels a thousand times lower. With no trend in time, the impacts are random.[1]

In the sea there are many more minnows than whales. In space too, collisions act to slowly grind down rocky material, so there are many more small rocks than large rocks, just as a beach has many more grains of sand or pebbles than large boulders. More small chunks of rock than large chunks of rock means there is an inverse relationship between the size of object and the frequency at which an object of that size hits Earth (Figure 6.1).

Let's start at the frequent but reassuringly puny end of the spectrum. A rock a meter across reaches the top of the atmosphere somewhere above Earth almost every hour. These objects almost all vaporize in the lower atmosphere; if you live near a dark sky you've probably

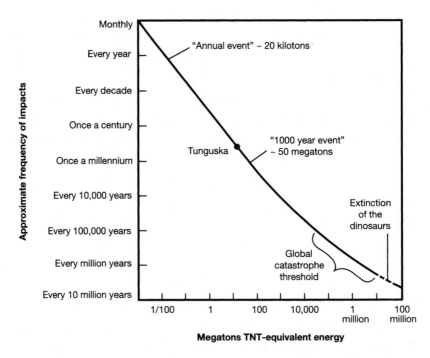

Figure 6.1. There are many more small chunks of space debris than large chunks and this produces an inverse logarithmic relationship between frequency of impacts and the size of incoming projectile. Earth's atmosphere quenches most of the smaller, frequent threats and the large threats are thankfully extremely rare.

seen a few of these as shooting stars. Projectiles five meters across arrive about once a month and release energy equal to the Nagasaki atomic bomb when they explode in the upper atmosphere. The public is unaware of these blasts but the U.S. Air Force tracks them with the satellites deployed to look for violations of the ban on testing nuclear weapons. Declassified documents show 136 atmospheric explosions between 1975 and 1992, in line with astronomical predictions. Before the military knew these events were cosmic, there was a danger they might have blamed the Russians, with disastrous implications.

On June 30, 1908, a remote area in Siberia was devastated by a huge blast eight kilometers above the ground. The forest nearby ignited, 80 million trees over 2600 square kilometers were flattened, and seismographs as far away as London registered the event. The Tunguska event packed a punch equal to the largest nuclear test ever, the Castle Bravo device. That's nine million metric tons of old-fashioned TNT. The incoming object was about 30 meters across, and such objects are rare enough that they hit Earth only about once a century. We were doubly lucky—in both time and space. If it had occurred elsewhere it might have killed many people and if it had occurred in the same place 50 years later, during the height of the cold war . . . well, you can imagine the rest.

The Barringer crater in northern Arizona is 50,000 years old but it's beautifully preserved by the arid desert climate. The meteorite was made of nickel and iron and it arrived at 13,000 kilometers per second (30,000 miles per hour), giving it great destructive force. A 50-meter missile left a mile-wide crater. Objects like this hit once a millennium, meaning they've left scars on human history. Barringer is five hours' drive from where I live; I've stood at the crater's center and marveled at the prodigious amount of solid rock displaced by the projectile.

An interdisciplinary team called the Holocene Impact Working Group has gathered evidence of an asteroid or large comet impact 4500 to 5000 years ago in the ocean east of Madagascar. Sediment deposits shaped like chevrons all point to a 30-kilometer crater in the Indian Ocean. Such an impact would have created megatsunamis 200 meters high and tree rings from the time show a chilled climate that could have been caused by the veils of dust thrown up by an impact. Detective work like this is controversial due to fragmentary evidence and the difficulty of ruling out other explanations.

Debate also rages over claims that one or more meteors slammed into North America 12,900 years ago, causing the extinction of mammoths, saber-tooth tigers, and an early human tool-making culture called the Clovis. Evidence includes tiny shocked rocks called tektites, an iridium layer, widespread carbon-rich charred soil, and tiny diamonds that can only be formed under extraordinary heat and pressure. But there's no crater, and an impact couldn't have been the sole cause of the end of the last ice age because the climate had been changing hundreds of years earlier. The onus remains on impact supporters to show that their explanation is compelling.[2]

Going up a level to the bigger fish, it gets ugly, fast. A 200-meter rock hits Earth about once every 100,000 years and deposits energy equal to 6000 megatons of TNT, more than the sum of the world's nuclear arsenal. The crater is eight kilometers across and the earthquake induced is 7.1 on the Richter scale. Even at a distance of 160 kilometers, there is damage from the blast wave and debris fragments. A direct hit on a city would kill millions. Remember, a 200-meter asteroid is a very modest-sized one. A one-kilometer asteroid is considered the threshold above which there are global effects from an impact.

When Light Shows Turn Deadly

The space junk described so far arrives unpredictably. That's why we can't take too much comfort from the fact that an impact of the size that killed the dinosaurs happens every hundred million years. Let's see, it was 65 million years ago, 100 minus 65 is 35, so we're not in any danger. That would only be true if the arrival was like clockwork every 100 million years. With random arrivals, the next "big one" is almost as likely to arrive on your next birthday as it is on the same day 35 million years from now.[3]

The timing of one type of impact is predictable, as far as knowing whether it will happen on your birthday. Meteor showers occur when Earth passes through debris strewn along the path of a comet that crosses Earth's orbit. The names of the showers are based on the constellations where the light show is concentrated. The Orionids that peak on October 21 are associated with Halley's Comet, the Perseids that peak on August 12 are associated with Comet 1862 III, and the Geminids that peak on December 14 are associated with the comet called 3200 Phaethon.[4]

The variability of meteor showers from year to year is a sign of the uneven distribution of debris along the comet's path. Chunks break away from the comet due to the action of the Sun and some can be large enough to potentially do damage. In a typical year the Leonids have a modest peak rate of about 10 light flashes per hour. But in 1833, Boston eyewitnesses described a tempest of falling stars at a rate comparable to snowflakes in a snowstorm, and in 1966, local eyewitnesses counted rates of 150,000 per hour over a 20-minute period, and sky illumination bright enough to wake people because they thought the Sun was rising.

The Taurids have a special role in human culture and consciousness. Often called the Halloween fireballs for their arrival in late October, Taurids are caused by debris from Comet Encke. This comet is unique in having an orbital period of only 40 months. Calculations suggest it entered the inner Solar System for the first time about 20,000 years ago, and since then has disintegrated into several chunks the size of asteroids, plus lots of substantial rubble. Encke's comet is a leftover from a much larger initial object. The Taurid "stream" has a cycle of activity peaking every 3000 years as Earth passes near its core.

Associations of cosmic phenomena with historical events are often very difficult to prove, but the Taurids have been implicated in a number of pivotal events in human history, most recently the Tunguska impact. There was also a peak at the time of Christ and when Stonehenge was first being constructed. Some scholars think the Bronze Age breakup of the original large comet caused destruction in the Fertile Crescent, supported by evidence in Iraq of a large meteor crater from that time. The event may even have found its way into the legend of Gilgamesh, where the contemporaneous account talks of the Seven Judges of Hell "raising their torches and lighting the land with flame" and a fearsome storm that "turned day into night and smashed the land like a cup."

Whether meteor showers have done more than fuel fear, wonder, and apocalyptic thought is open to question. It's easy to imagine that the recurrence of meteor showers was an early token of human culture, one of our first "folk memories." The major showers put Earth in the firing line with orbital regularity, increasing the odds that, one year, the light show will turn deadly.

Saving the Planet

Some Close Calls

Study of space debris has come a long way since Thomas Jefferson passed judgment on a report that stones had fallen from the sky by saying, "I would more easily believe that two Yankee professors would lie than that stones would fall from heaven." To give him his due, he was later convinced by the evidence and used the meteorite example to argue for the power of the scientific method. The lack of unaltered surfaces means that craters are rare on Earth and even as late as the 1950s landmarks like northern Arizona's Barringer crater and the circular features on the Moon were being attributed to volcanism.

Over a hundred large craters have been found, supported in many cases by evidence of material of extraterrestrial origin (Figure 6.2). Even the impact rate isn't entirely based on speculation. *Mars Global Surveyor* directly measured the current impact rate of small objects when it observed 20 fresh craters as its camera scanned the planet between May 1999 and March 2006.

Apart from the notorious incidents that have entered the pop culture—Ann Hodges grazed in the thigh while she lay on the sofa in her living room, Michelle Knapp's Chevy Malibu being hit inches from its fuel tank, the Egyptian dog that was purportedly taken out by a rock from Mars—we've had some close calls. In August 1972, many

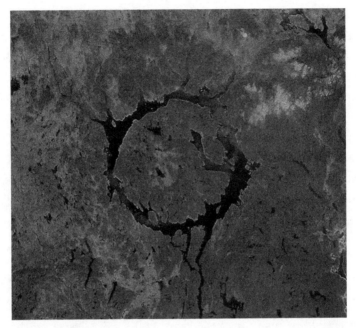

Figure 6.2. The Manicouagan crater in Quebec is 200 million years old and 70 kilometers wide, one of the oldest and largest craters known. Worn away by glaciers and erosion, but clearly visible, it was photographed by the space shuttle *Columbia* in 1983.

people in the northern Rockies witnessed a daytime fireball caused by an object the size of a house that grazed the upper atmosphere and headed off into space, like a stone skipping off a water surface. In May 1996, a 500-meter asteroid was detected and a few days later passed by 450,000 kilometers from Earth, about twice the Earth-Moon distance. In March 2004, a smaller 30-meter asteroid passed only 40,000 kilometers from Earth.

It's enough to give you the heebie-jeebies. I'm not prone to obsessive thinking, but every now and then a mental image seeps in and is hard to shake off. In Thomas Pynchon's sprawling novel *Gravity's Rainbow* the hero is an American serviceman living in London during World War II, at a time when German V-2 bombs are raining in randomly on the city. He's haunted by the idea that a V-2 bomb, arriving supersonically, might be falling directly toward his head as he walks down the street. That's the way it would be for an incoming meteor.

What's a cautious person to do? First, avoid open spaces. Working and living on lower floors of tall buildings is another good strategy; let the penthouse dwellers take the hits. Flying is a bad idea because you're above much of Earth's protective blanket. Don't go out after midnight when Earth's orbital motion combines with its spin to increase the chances a projectile will make it to the ground. Sleep standing up, by hanging in a bag in a closet, for example. Your cross-sectional area to meteors is much smaller when you do. Ann Hodges's mistake was to lie down, just asking for trouble. Last, banish these obsessive thoughts before your friends start worrying about you. Astronomer Alan Harris has calculated that the lifetime odds of dying by an impact are 1 in 700,000. You're more likely to die in a terrorist attack and you don't worry about that, do you? OK, bad example.

Be assured that astronomers are keeping a watchful eye on the threat, though they're of no use for spotting the small stuff. NASA funds several teams to scan the skies with small telescopes; their goal is to detect and measure orbits for 90 percent of the objects one kilometer or larger with potentially Earth-crossing orbits. That might leave you worried about the 10 percent that get away or the imperfect ability to find objects several hundred meters across, which pack a big wallop. But NASA's Web site is reassuring. Sentry is the name of the automated collision monitoring system and in December 2008 the highest impact probability in the list was 1 in 3000 for an object 130 meters across. Nothing at the moment has a color threat of yellow or worse on the international "Torino" scale that was adopted by researchers in 1999.

We had a ringside seat, with popcorn and colorful images provided, when Comet Shoemaker-Levy 9 collided with Jupiter in 1994. This comet had already been ripped apart by Jupiter's gravity when the fragments slammed into the gas giant's upper atmosphere. Twenty-one impacts were seen over six days, the largest caused by a two-kilometer chunk that left a mark larger than Earth and deposited 600 times more energy than the world's nuclear arsenals. This event lodged the possibility of impacts firmly in the public consciousness.

It's fitting that the comet was codiscovered by Eugene Shoemaker. His 1960 PhD convincingly showed that the Barringer crater was an impact and he went on to found the field of astrogeology. Shoemaker had hoped to walk on the Moon but Addison's disease ruled him out of an Apollo mission so he settled for training the astronauts. In a bitter irony, the man who worked his whole life on impacts died in a head-on collision north of Alice Springs in Australia. The Lunar Prospector mission took some of his ashes to the Moon; he is the only person to be honored with burial on another astronomical body.

The Big One

Hollywood rolls out a space disaster movie every decade, and they're not generally Tinseltown's finest efforts. *When Worlds Collide* in 1951 had Earth threatened by a planet and a star, and *Meteor* in 1979 got the Soviets and the Americans cooperating to destroy a five-kilometer inbound asteroid. In 1998 not one but two mediocre movies revolved around impact catastrophes: *Deep Impact* and *Armageddon*. The first has a few merits but bad acting trumps bad science in both.

Here's the real science of the "Big One," and it's just as scary as any movie. You can explore the possibilities yourself thanks to Jay Melosh and his colleagues at the University of Arizona. He's created an online Catastrophe Calculator, where you plug in the size, composition, and incoming speed of the projectile, pick your distance from ground zero, and dial up the very scary numbers.[5] "If you're close to the site of a major impact, some pretty bad things happen," Melosh warns, with laconic understatement.

Let's say it goes down in Los Angeles. Nothing against the City of Angels but for dramatic effect we'll liken it to modern-day Gomorrah. You're standing on a hill in Daly City, south of San Francisco, looking toward Los Angeles 563 kilometers away. The 10-kilometer bolide streaks across the sky at 30 miles per second. A staggering equivalent of 4.5 trillion metric tons of TNT energy is liberated as it hits. Downtown LA is leveled instantly. Three thousand cubic miles of rock are vaporized and flung into the air. The fireball over what used to be LA is 100 times larger and 1000 times brighter than the Sun. When the crater settles it will stretch from Oxnard to Riverside, and the San Bernardino and San Fernando valleys will be covered 1600 meters deep with magma.

You'd need to be in a battle-hardened silo in Daly City to survive what happens next. Five seconds after impact the thermal radiation arrives, igniting grass and trees and melting flesh and paint. The earthquake reaches San Francisco after two minutes. It's Richter 10.4, or greater than any in recorded history. Even five or six hours' drive away, many buildings suffer moderate to severe damage.

Six minutes after impact centimeter-sized debris fragments arrive, traveling at high velocity. By the time they've all arrived, northern California will be covered with an ash and rock blanket five meters thick. The final punch comes from a blast wave that arrives after another half an hour. It's traveling at hundreds of kilometers per second and it's a lot louder than the front row seat at a rock concert. The blast wave levels most buildings and 90 percent of all trees, knocks down bridges, and tosses cars and trucks like chaff on the wind. Are we having fun yet?

This is an extraordinary impact, but only a bit larger than the one that killed the dinosaurs. Jay Melosh, the "Grim Reaper" of impacts, spares the visitor to his Web site the more graphic details. "We could have put in some worse stuff," he says genially, "but it started getting grisly."

And What to Do About It

An impact like the one just described has happened seven or eight times in the history of the Earth. Killer asteroids or comets will come—it's certain. All that's uncertain is the timing. Even if the odds are extremely low in a human lifetime, it's worth having a plan.

The first step is vigilance and detection. An international network of small telescopes scans the skies every night for objects that move in the fixed pattern of stars. A dozen or so observations are enough to measure a crude orbit and decide if it's Earth-crossing. Over 90 percent of the near-Earth asteroids have been discovered since 1990. About 800 are larger than a kilometer and that number grows slowly because most large asteroids have been found. The total number is over 5000 but is growing rapidly because there are so many small asteroids. It's relatively cheap to fund this work; it requires only small telescopes.

One concern is the need to detect short-period comets in the inner Solar System. If they're old their ices will all have boiled off so they don't light up like normal comets when they're near the Sun. They orbit through a spherical region, which means they can be found in any direction in the sky, unlike asteroids, which are always near the plane of the ecliptic. That makes comets harder to find.

The business of cataloging the threat took a big step forward in 2008, with the commissioning of the Panoramic Survey Telescope and Rapid Response System in Hawaii. Each of these four telescopes will have a camera with 200 times the pixels of a typical digital camera, making it able to scan the entire night sky every week. The project goal is to complete the census of near-Earth objects larger than a kilometer and catalog as many as possible of the objects larger than 300 meters.

The Large Synoptic Survey Telescope is even more ambitious. This 8.4-meter telescope will survey the visible sky down to a level 100 million times fainter than the eye can see *every week*. Such impressive light grasp is possible thanks to a camera slightly more eye-catching than a digital camera you might slip into your pocket; it weighs nearly three metric tons, is the size of a small car, and has 3200 megapixels. The exquisite mirror for the telescope has already been cast at the observatory where I work—its largest imperfections are like bumps an inch high on a surface the size of the continental U.S.—and the telescope itself is due to start operations in Chile in 2014 (Figure 6.3). It will be able to locate 90 percent of the near-Earth objects larger than 140 meters, fulfilling a mandate from Congress to identify all potential threats by the year 2020 and let us sleep a little more soundly. Then.

The second step is calculation. Cheap, fast computers have made it easier to make models of orbits and project them forward in time. NASA's Sentry program calculates the probability of an impact in the next 100 years. Beyond that, the orbits are too uncertain for reliable predictions. A lot of people got a scare back in 2004 when they read that the hefty asteroid 99942 Apophis, 300 meters across, had a 1 in 37

Figure 6.3. The mirror of the Large Synoptic Survey Telescope is 8.4 meters in diameter and it will be able to gather enough light to detect space rocks 160 meters across, a billion kilometers away. The telescope is due to start operations in Chile in 2014. The mirror was cast under the football stadium at the University of Arizona.

chance of hitting us on April 13, 2029. Later observations showed that it will miss Earth by a slender 26,000 kilometers in 2029, within the altitude of communications satellites, and its odds of an impact in the twenty-first century are only 1 in 45,000.

It's often the case that the risk level of collision recedes as more data are taken. Is this NASA crying wolf? No, the early orbits are uncertain so they "permit" more outcomes that include a future impact. With more data the predictions sharpen and it's much more likely that an asteroid will follow one of the many trajectories that doesn't lead to a collision as opposed to one of the tiny number that does. For a well-studied asteroid the limitation on prediction stems not from the measured orbit, but from the properties that can't be measured easily that affect the orbit, like spin, reflectivity, and objects it might interact with in the future.

If an object stays on the danger list as the number of observations increases, it's time to get out of the recliner and do something. Here's where Hollywood gets it badly wrong. It sounds great to dust off the nukes and blast the sucker to smithereens. But the smithereens are moving on the same trajectory and could collectively cause even more damage than the original object.

Beyond the cowboy approach there are many strategies, some quite ingenious. At a Planetary Defense Conference in 2004, people from NASA and the aerospace industry considered the alternatives. One of the session chairs was Rusty Schweickart, a former astronaut whose B612 Foundation has the goal of testing the technology to sidetrack an asteroid by 2015. The foundation's Web site is an obsessive compendium of risks and remedies regarding impacts. It should include a motto: We worry so you don't need to. "If we do not prevent such an occurrence when we have the capability to do so," says Schweickart, talking about a devastating impact, "it would be the greatest crime in human history."

One idea is to position a spacecraft alongside the asteroid and use it as a gravity "tractor" to slowly pull it onto a different trajectory. We could also put reflective materials on it and use radiation pressure from the Sun to push it sideways slightly. Or we could use a large reflector to focus sunlight on it and create a hot jet of material that would act like a little thruster. More conventionally, we could attach a jet propulsion system to it to deflect it onto a safer path (Figure 6.4). None of these ideas is far-fetched because we've sent spacecraft to comets before, but they'd each take a decade or more to implement, so we can't breathe easy just yet.[6]

Life Is Viral

Sterilizing the World

We've see that the biosphere is pervasive, robust, and durable. Life on Earth started very early and radiated into every conceivable ecological niche. Tracers of biology stretch continuously back for nearly 4 billion years. In thinking about the end of the biosphere the right question to ask is: what would it take to sterilize the world?

Figure 6.4. This is an artist's impression of the encounter between *Deep Impact* and Comet Tempel 1 in July 2005. The mission sent a projectile five times the mass of a person crashing into the surface with the aim of seeing what a comet is made of, but it was also a test of how well we could approach and alter the trajectory of a comet or asteroid if Earth were threatened.

Biology has suffered indignities that were not impacts; only one of the mass extinctions has been explicitly tied to a crater of the appropriate size and age. Other mass extinctions were probably caused by intense epochs of volcanism. The Permian extinction 250 million years ago—called the "Great Dying" because 95 percent of species disappeared—was most likely caused by a volcanic event associated with the breakup of the primal supercontinent, Pangea. An area the size of the United States was coated by several hundred meters of lava in the blink of a geological eye, and a large amount of oxygen was removed from the oceans. At least twice, coupled changes in volcanism and atmospheric physics caused the oceans to mostly freeze over in a Snowball Earth.

No geological event is violent enough to erase all life so only an impact can kill the biosphere. Calculations by Norman Sleep, a geophysicist at Stanford University and other colleagues, suggest that a 400-kilometer asteroid would vaporize the oceans and blanket Earth with a shroud of 1100°C (2000°F) molten rock droplets that would rain down and create a layer of rock 300 meters deep. The oceans would take 2000 years to condense from steam. This is very probably a sterilizing event.[7]

An impact of half the size, 200 kilometers, would vaporize the oceans and create enormous volumes of molten rock, but the adverse effects would only last for 300 years. At 100 kilometers, there might be deep ocean pockets that are not evaporated, so this marks the boundary of survivability. Impacts of this size probably occurred during the heavy bombardment 4.1 to 3.8 billion years ago but there's no evidence that an asteroid or comet larger than 30 kilometers has struck Earth since.

What about the future? If we say 100 kilometers is the minimum sterilizing size, the average time between such impacts is about 20 billion years. So it's unlikely to happen for billions of years, although the random nature of the timing means there's a small probability it could arrive a lot sooner. However, 100-kilometer asteroids are easy to keep track of, so we'd have plenty of warning.

Taking Over the World

Biodiversity spans bacteria to blue whales and it's hard to imagine life without a cornucopia of species. But what if the fate of the biosphere were to revert to just one thing: microbes? Recall that the world was exclusively microbial until the Cambrian explosion. Animals have 10 times more microbes than cells, so microbes already own the world and use our bodies as hotels. The interdependence of the web of life seems reassuring; viruses can't live without cells and bacteria occupy evolutionary niches defined by multicelled plants and animals.

Consider then the deeper story of the Permian mass extinction 250 million years ago. It's not true that "all life" nearly died—the microbes sailed through unscathed. In fact, there's biomarker evidence that the microbes contributed to the extinction and were its beneficiaries. Note that this theory is still controversial and will need additional evidence before it's widely accepted.[8]

Here's what geochemists think happened. The trigger was a volcanic eruption of unprecedented scale that created the Siberian Traps flood basalt. The rapid increase in carbon dioxide warmed the world enough that the temperature difference between poles and equator was small, causing the air and ocean currents that keep the oceans oxygenated to be quenched. As a result, the types of green and purple bacteria that get their energy from photosynthesis using hydrogen sulfide took over. The oceans filled with toxic gas, killing most other organisms. Then the "rotten egg" gas escaped into the atmosphere, killing almost all plants and animals as well as depleting the ozone layer that protects Earth from harmful ultraviolet (UV) radiation. Bacteria poisoned the planet.

Then they completed their takeover. Rocks from the late Permian and early Triassic show that diverse coral reef fossils gave way to simpler sediments made of bacterial mats. The seafloor became coated with microbes. It took many millions of years for multicelled organisms to claw their way back into the ecosystem.

If it happened once, it could happen again. We shouldn't assume that large creatures like us are the end point or necessary consequence of evolution. The diversity and adaptability of microbes will always give them the edge when conditions change radically. If it happens in the course of natural selection we'll have time to react and maybe counter the problem or, in the worst-case scenario, find a safe haven off Earth or underground. But there's a worrying scenario where we engineer the same outcome inadvertently.

It's called the "green goo" problem. Gray goo is a standard plot line in science fiction, where self-replicating nanobots spread uncontrollably and take over the world. Green goo is a version of this scenario where genetic engineering and nanotechnology combine to create a "perfect" microbe, one that outcompetes existing microbes and takes over the biosphere. Mergers of living and nonliving matter are likely to behave in unpredictable and possibly uncontrollable ways. Think tanks have produced reports on green goo and the Swiss Re, the sober company that insures insurance companies, has weighed in on the risks. At the very least, it's a possibility we should keep our eye on as we develop our biotechnology future.

Eternal Biospheres

Earth is a living rock hurtling through space. Creatures in the surface biosphere are protected from the rigors of the external environment by an atmosphere that regulates the temperature and keeps out harmful UV radiation and cosmic rays. If life can travel through space on large rock, maybe it can travel on a small rock? What if life is so viral that it has "infected" other worlds?

Panspermia is the idea that the "seeds" of life exist elsewhere in the universe, and that life on Earth and other planets might have gotten its start from these seeds. Another term for this is exogenesis. In the nineteenth century Lord Kelvin was thinking about the implications of Pasteur and Darwin's work. Pasteur had showed that life comes from life and does not spontaneously arise. Darwin had showed that one form of life can evolve into another but he resisted speculating on the origin. It was to avoid the problem of origination that Kelvin suggested that organisms could hitchhike on rocks to cross the space between planets and stars.

The idea was further developed by Svante Arrhenius, the founder of physical chemistry and the originator of the theory of the greenhouse effect. Arrhenius argued that microbes could travel through space as dormant spores, driven by the Sun's radiation pressure. In the 1970s Fred Hoyle and his student and collaborator Chandra Wickramasinghe gave new life to panspermia by claiming that interstellar dust rains down on Earth and other planets to jump-start life. Their arguments failed to convince most researchers. They claimed that comet nuclei and interstellar dust contain viruses, while must astronomers see no evidence for more than the basic building blocks of life in space. And they claimed that the patterns of flu epidemics and even outbreaks of SARS were consistent with an extraterrestrial origin for the viruses, something the epidemiology community finds very unlikely.

Panspermia doesn't solve the problem of the origin of life; it passes the buck to an undisclosed location beyond Earth. However, it does relax the requirements in two ways. First, it allows for multiple sites and mul-

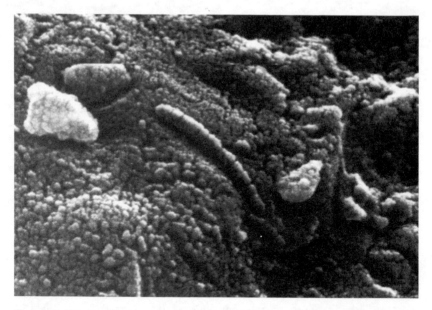

Figure 6.5. This famous photo from an electron microscope was the most evocative image of the meteorite Allan Hills 84001 when a NASA team claimed that it had found traces of life in it in 1996. The elongated structure is much smaller than any cellular forms on Earth, and similar morphologies can result from rock "splatter" during impacts. It's also possible that the rock was contaminated by terrestrial biology.

tiple originations on a wide range of environments beyond Earth. Second, it expands the time frame for life to start, from 300 million years between the end of heavy bombardment and the first secure evidence for microbes, to 9 billion years between the earliest carbon and planets in the Milky Way to the first microbes on Earth. The extraordinary durability of extremophiles also makes the premise of panspermia seem plausible.

There are two types of panspermia. The first involves the transport of living organisms within the Solar System. Impacts on the surface of a planet or moon can eject small rocks into space—think of what would happen to a rock on a trampoline if you jumped on the trampoline. As shuttle systems go, it's very inefficient, but we know material can get from Mars to Earth because several dozen Mars meteorites have been found and it's estimated that one Mars rock falls to Earth each month. The situation is favorable for Mars to Earth travel because the gravity of Mars is weak and its atmosphere is thin, it's the closest planet, and the energy arguments favor motion toward the Sun.

The claim of life traces in the Allan Hills meteorite in 1995 raised the possibility that we are all Martians (Figure 6.5). It turns out that the evidence for life was either overstated or ambiguous, and most space scientists hold that the case for biological traces is unproven,[9] but the possibility should be taken seriously. Mars is a small planet that cooled quickly and was hospitable for life even more quickly than Earth. Three billion years ago it had a thick atmosphere and was warm and wet.

Tests with space guns at U.S. weapons labs have shown that bacteria can survive the thousands of g's experienced when a rock is blasted off the surface of a planet. A few centimeters of rock will shield life from harmful cosmic rays and UV rays in the vacuum of space. That only leaves one obstacle: the fiery re-entry.

In 2007, a team from the European Space Agency strapped two small sedimentary rocks to the heat shield of an unmanned Foton capsule. On its return the capsule entered the atmosphere at 8 kilometers per second (18,000 miles per hour) and the heat shield reached a tem-

Figure 6.6. In 2007, the European Space Agency tested the survival of life and evidence of life in the harsh conditions of space by using two samples of sedimentary rocks and a control sample of basalt attached to the heat shield of a Russian Photom M3 rocket. The basalt was lost in space but a sample of 3.5-billion-year-old volcanic sand containing microfossils and a 370-million-year-old Orkney mudstone containing chemical biomarkers survived with their ancient fossil evidence intact.

perature of 1700°C (3090°F) (Figure 6.6). The bacteria smeared onto the surface of the rocks were vaporized, but chemical traces of life and microfossils a few centimeters deep within the two rocks were unaltered. Other tests have shown that a potato-sized meteorite will stay cool at the center on re-entry. The Planetary Society plans a more comprehensive experiment in 2011 when they launch a biological package the size of a hockey puck on the Russian Phobos-Grunt spacecraft. It will house samples from all three domains of life: bacteria, eukaryote, and Archaea. After three years in deep space, the samples will be returned in 2014.

The second type of panspermia is transport between stellar systems. The nearby stars are millions of times farther away than Mars is from Earth, so a space rock would have to make a much longer and lonelier journey. Jay Melosh, the impact "scare-monger" we met earlier, has calculated the orbits and odds. Even though billions of rocks have been ejected from the terrestrial planets over 4 billion years, Jupiter is a very inefficient slingshot—meaning most don't leave the Solar System. After fanning out over vast interstellar distances, the odds are that only a handful of rocks has ever been captured by any other star, and none has ever hit the surface of a distant terrestrial planet.

Whatever the fate of our biosphere, there's no natural way that life can spread through the Galaxy. Life on Earth shares the commonality of its genetic material; it's all one thing. Life in the Milky Way will be as diverse and distinctive as the environments that give rise to it.

Chapter 7

LIVING IN A SOLAR SYSTEM

The rock is not much to look at—pale gray, grainy, and sedimentary, no bigger than a potato. It's one of thousands liberated from the weak tether of Martian gravity when a meteor hits the salty sea of Meridiani Planum. Like many of the ejected rocks it contains samples of the dense microbial mats that cover the floor of the shallow sea. So soon after the Solar System forms, debris litters the space between planets and impacts are common.

Time passes. Most of the rocks ejected from Mars travel endlessly in deep space. This rock is special because its path in space and time intercept the location of Earth a million orbits later. It falls out of a blue sky onto a sterile planet; Earth has been devastated by a series of huge impacts that vaporized the oceans and melted the rocky surface, obliterating the primitive microbes that had formed in the planet's Hadean era.

The rock heats up as it plunges through the thick atmosphere but the center remains cool. Years after it lands, cycles of freezing and thawing split it open and rain dissolves some of its interior and washes the material into a shallow pond. The microbes rehydrate and reanimate. The local chemicals present a challenging environment, but after thousands of generations of adaptation the microbes thrive in the new conditions and proliferate into many new ecological niches.

The paths of the home planet and the adopted planet diverge. Mars does not have enough gravity to hold a thick atmosphere or drive tectonics, so it dries out to become a frigid desert. Earth has a thick atmosphere flushed with the respiratory by-product of photosynthesis, and in time a tree of life emerges, with byzantine biological diversity.

Many, many orbits later four of Earth's larger organisms are ejected from the planet. They travel not in a rock but in a metal canister containing a microcosm of the planet's atmosphere. As they watch anxiously through a small window, Mars swims into view. Later, on the surface, they work with purpose and move easily in the gentle gravity. They are very far from home. They have come home.

Pale Blue Dots

Rare Earth

So far we've considered endings within the context of three familiar and nested realms: the human organism, the human species, and the biosphere that contains all humans. Now we step outside the world of the familiar into the world of the cosmic and unfamiliar.

There's a basis for understanding the end of our lives because billions of people have died through history and, similarly, there's a basis for understanding the end of our species because so far all species have been subject to natural selection. But with the biosphere and Earth we encounter the problem of having only one example to study and learn from. For signposts as to how the world will end we need to find other examples of living and Earth-like planets. Without extra information, there are two different ways to interpret the fact that we're intelligent beings on a long-lived planet that hosts abundant life.

Earth might be rare. Peter Ward and Don Brownlee from the University of Washington provoked vigorous discussion in academic circles and beyond with their popular book *Rare Earth* in 2000. The authors agree with most scientists that the range of extremophiles on Earth implies that microbial life may be fairly common on terrestrial planets beyond the Solar System. But they make a strong distinction between life of any kind and "complex" life, which for this purpose can be defined as large creatures with intelligence. They don't have to have technology, and they don't have to be humans or primates, but they're large and sophisticated multicelled creatures with brains or something similar.

Ward and Brownlee argue that the evolution of complex life needs a long-lived, stable environment and a set of fairly special conditions. These include a nearly circular orbit around a long-lived star like the Sun, a suitably "quiet" environment in the Milky Way, the presence of a giant planet like Jupiter to protect against impacts, a large moon to stabilize the orbit, sufficient water, and plate tectonics. The list is intimidating and the Rare Earth argument centers on a supposition that all of these attributes are both unlikely *and* essential to evolve complex life.

The last decade has eroded much of the attraction of these arguments. For example, simulations show that water delivery and plate tectonics are going to be normal attributes of rocky planets within a few Earth-Sun distances of their stars, and the presence of a Jupiter-like giant can cause as many giant impacts as it prevents. More fundamentally, it's not been proven—and might be impossible to prove—that any of the attributes are *essential* for complex life. We've seen that life and the environment exist symbiotically so it's a circular argument to use the attributes of the physical environment to argue that a particular outcome is unlikely. All of which points us in the opposite position.

Earth might not be rare. Astronomy has operated successfully under a Copernican principle or "principle of mediocrity" for the past 400 years. It's a heuristic or educated guess rather than a formal theory. At every turn, as we learn more about the universe we find we're not privileged or unique. The Milky Way, our position within it, and the star we orbit aren't unusual or special. Life's core ingredients—carbon and water—are universally created and distributed. And planets beyond the Solar System are being found almost weekly; the current census stands at nearly 400 and the detection limits will soon include Earth clones.[1]

Think of planets as grains of sand. The number of grains of sand in a square meter patch of beach, assuming the sand is a meter deep, is about 10 billion. If we ignore the gas giants and their moons, that's a plausible estimate of the number of terrestrial planets in the Milky Way galaxy. Now imagine examining each grain of sand in that cubic meter. In the Rare Earth scenario, you would be very unlikely to encounter any grain of sand that's habitable in the way that our grain of sand is habitable. But if there's nothing special about our situation, the cubic meter may contain thousands or even millions of Earths.

The Rare Earth hypothesis includes time. It took 4 billion years on this planet to go from slime to civilization (Figure 7.1). Long spans of uninterrupted evolution aren't possible in many regions of the Galaxy and around many massive stars. How do we decide if the time it took for complex life to develop on Earth makes it likely or unlikely to have developed elsewhere?

Princeton astrophysicist Richard Gott thinks the Copernican argument can be applied to situations where you only have one sample, as long as you have no reason to believe you're observing the situation at a special time. The logic is so simple it seems it must be wrong. You're more likely to observe something in the middle of its existence than at the beginning or the end. There's a 50 percent chance of observing something during the middle half of its existence and a 95 percent chance of observing it outside the first 2.5 percent and last 2.5 percent of its existence, and so on for any window of observation.

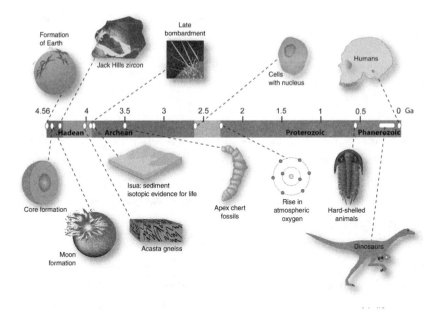

Figure 7.1. In a timeline of the history of Earth, our position as intelligent creatures is very recent. Statistical arguments can be made that assign a longevity to humanity based on the fact that our location in history is not special. On the other hand, we have no general theory for the evolution of intelligence.

Gott has applied this reasoning to world leaders and dogs and Broadway musicals but let's use his example of humanity. We've been around for about 200,000 years. By the Copernican timing argument, 200,000 years is unlikely to be in the first 2.5 percent or the last 2.5 percent of humanity's entire span. With 95 percent probability, the human race will survive at least 5100 years and less than 7.8 million years, about what we might predict by looking at other species of mammals.[2]

We can stretch the Copernican idea to the timing of emergence rather than longevity. Suppose complex life exists elsewhere in the universe and the time it took to emerge on Earth isn't special in any way but is representative of the overall distribution. It took 4 billion years for complex life to evolve on Earth. With 95 percent confidence, we say that it should take less than 800 billion years and more than 100 million years anywhere else. The range is huge, but informative. At the high end, only the lowest-mass stars could host life and they have slender habitable zones. At the low end, even Peter Ward concedes it might only take 100 million years to develop complex life, which would bring into play high-mass stars with much shorter lifetimes than the Sun.

Finding the Perfect Marble

When you were a child somebody gave you a marble. It was so long ago you don't quite remember who or when. The marble is beautiful; it has swirls of blue and green on a white background. You carry it around with you and are quite attached to it. This marble is your prized possession.

Let's say this marble is a terrestrial planet with complex life: a world like Earth. If this were the only marble we'd ever seen, or ever known, how much would we be justified in concluding about the properties of all the other marbles, or planets, out there? You'd probably say, not much or nothing, and yet scientists on either side of the Rare Earth debate are trying to argue that the marble is exceptional or typical!

We're stuck with a basic problem of induction: generalizing from a sample of one. That's logically hazardous and without some general "theory of marbles" could lead to an erroneous conclusion. Suppose there is a large black velvet bag of marbles, representing all of the planets beyond the Solar System. We're allowed to draw one marble from the bag. We do, and it's beautiful—a white marble swirled with blue and green. What can we conclude about all the marbles in the bag? We could speculate that they're all like the first marble, or some of them are, or very few of them are. But in truth, we can say very little except that at least one white marble with blue and green swirls was in the bag and we were fortunate with our first pick.

Now suppose there are three kinds of marbles in the bag. There are large black ones that represent planets like Jupiter, gas giants with no life. There are small white ones that represent terrestrial planets like Earth, with white indicating microbial life. And there are white ones with extra colors like our first marble, representing terrestrial planets like Earth that have evolved complex life. By setting up the situation this way, we've included the assumption that all terrestrial planets harbor life of some kind, but we've said nothing about the relative numbers. Each type could be common or rare.

If we draw one marble from the bag and it's the white kind with blue and green, we're still none the wiser. But if the first three are all like that we might start to suspect such marbles are common, and if the first 10 are all like that our confidence would increase. On the other hand, if only 1 of the first 10 marbles is white with blue and green, eight are plain white, and one is large and black, we could tentatively conclude that terrestrial planets are common but complex life is quite rare.

Unfortunately, at this stage of the search for extrasolar planets, the situation is quite different. Imagine you can't reach into the bag but have to select planets with a ring attached to a long handle. The ring catches the big black marbles but the small white ones all slip through. Marble after marble comes up large and black. Even though you are convinced there are smaller marbles in the bag, you can't snare them (Figure 7.2). Astronomers are easily detecting planets like Jupiter and Uranus, but they can't routinely detect planets like Earth and Mars yet. All conjecture on the abundance of white marbles of any kind is highly uncertain. But the rapidly rising mass distribution toward the limits of current surveys implies that there's a huge "iceberg" of lower mass planets awaiting discovery.

On Being Special

If discussing the Rare Earth hypothesis, it's common for the ideas of scarceness and specialness to be conflated, so let's try to keep them separate. Planets like Earth may or may not be rare; that's something that observations can and will decide in due course.

It's an entirely different question whether or not planets like Earth are the only planets where complex life can evolve. If our ideas of complex life and how it evolves are too Earth-centric then we may have done nothing more than invent a "just so" story to explain why the world around us had to be the way it is for us to be here. We're still at the stage of struggling to even detect Earth-like planets. Characterizing them well enough to say whether they could be or are habitable is a few

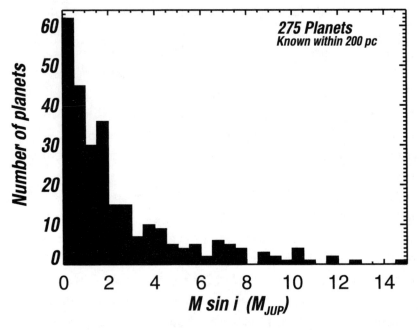

Figure 7.2. The distribution of mass of more than two-thirds of the exoplanets discovered so far out to a distance of 200 parsecs or 650 light-years. The inclination of the system affects the proportion of the full Doppler effect detected, and that inclination is usually unknown, so the x-axis is the product of the mass and the sine of the angle of inclination (in other words, the mass is a lower limit). The distribution rises rapidly near a Jupiter-mass, which until recently was the limit of most surveys. This implies a large number of undiscovered lower-mass planets near the left-hand edge of the graph; the current view is "censored" by the limits of detection.

years away and actually detecting signs of life on distant Earths may be 10 to 20 years off.

David Grinspoon is a planetary scientist and part-time rock guitarist. He's a hands-on scientist who settles for learning from space probes, but yearns to roam the surface of strange worlds. He has an unusual perspective on what constitutes a special outcome: "My cat Wookie survived life as a near-starving alley cat and wound up as a beloved house cat through an unlikely series of biographical accidents. . . . Trust me, given all of the incredible things that had to happen in just the right way, it is much more likely that there would be no Wookie than Wookie. I do not conclude from this that there are no other cats (The Rare Cat Hypothesis), only that there are no other cats exactly like Wookie."[3]

Our concept of "specialness" may not be useful. We only know one of the pathways by which life evolved to become intelligent; it might be the only way but we have no justification for that conclusion. It's more likely that our thinking or evolution is constrained by our environment and lacking in imagination. Jack Cohen and Ian Stewart wrote a book called *What Does a Martian Look Like?* which was in part a rebuttal of the Rare Earth concept.

Just as extremophiles thrive in diverse physical conditions that are normal and unexceptional to them but extreme to us, it's possible that more advanced or complex life can also thrive in unfamiliar conditions. However, a day at the beach to us might be horrific to such creatures, or as Cohen and Stewart describe it: "sleeting electromagnetic radiation, a corrosive oxygen atmosphere, and that terrible universal solvent, hydrogen monoxide, sloshing all over the place."

Let's return to the Copernican argument that our moment in history isn't special. Australian physicist Brandon Carter presented a first form of the argument developed later in great detail by Princeton astrophysicist Richard Gott. Assume only that we live neither among the first few or last few members of our species—that our place in the "roll call" of humanity isn't unusual or privileged. The number of humans who have ever lived is about 100 billion; that's a good estimate of our number in the roll call.[4]

Now consider two options for the future. The pessimistic view is that humans only survive another couple of centuries, so the total number of humans who will ever exist is 120 billion. The optimistic view might be one where humanity survives many millennia, perhaps even going off-Earth and being fruitful and multiplying, so that many trillions of people are destined to be born in the future. According to Carter, the "principle of mediocrity" argues that we should believe the pessimistic scenario, since our place in the roll call would be about 80 percent in, which is typical and unsurprising, whereas in the optimistic scenario we would be very early in the total roll call of humanity. Let's not get too comfortable about our destiny.[5]

Life Beyond Earth

Venus and Mars are Alright

The travelers encounter a solar system with three terrestrial planets. One is a close twin of Earth's but the atmosphere has nitrogen, no oxygen, and a little bit of carbon dioxide. Its single supercontinent is surrounded by a salty, mineral-rich sea. The second has a thick soupy atmosphere of carbon dioxide, active volcanoes, and numerous geysers that erupt into warm rivers in the highland areas. The third is smaller, with a tenuous atmosphere and shallow lakes dotting the surface. The melting from polar ice caps creates a dendritic network of waterways. There's an abundance of microbial life on each of these planets.

The solar system just described is ours 3 billion years ago. The first planet is Earth, the second is Venus, and the third is Mars. The only speculation is the last sentence—Venus and Mars were substantially more hospitable for life 3 billion years ago than they are now but we don't know if they were actually alive (Figure 7.3).

Recent evidence points in the direction of Venus being more habitable than Mars at some points in its history. Mars is now dry and cold, and the *Mars Reconnaissance Orbiter* has thrown cold water on the more tantalizing evidence for recent run-off and shallow sedimentary seabeds. On the other hand, models of Venus suggest it kept liquid water oceans for a billion years after formation, giving lots of time for life to form after the end of heavy bombardment. David Grinspoon is keen on sending a lander to Venus to look for traces of past life on small parts of the surface that haven't been resurfaced by volcanism. It would be a challenging mission given the forbidding physical conditions. He even thinks that life might have migrated to an ecological niche in the dense layers of cloud after the oceans evaporated. Mischievously, he points out that there might just as easily be Venusians as Martians.

Earth is not in the fat sweet spot as a Goldilocks planet. It clings to the edge of habitability. Simulations of terrestrial planets ranging from

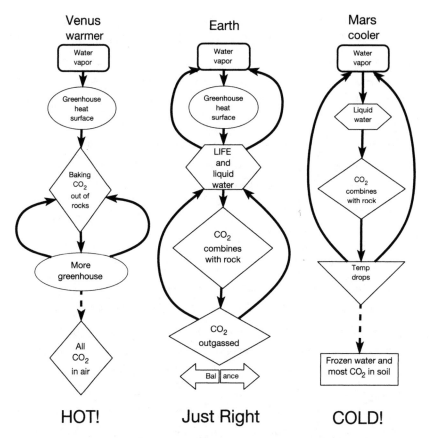

Figure 7.3. A flowchart showing how three terrestrial planets in the Solar System that were probably all habitable 3 billion years ago evolved in different ways. Venus was victim of a runaway greenhouse effect, Mars dried out and lost most of its atmosphere, and only Earth stayed habitable on its surface.

1/2 to 10 times Earth's mass show that plate tectonics can't operate when a planet is much smaller than ours. Plate tectonics is a prime, perhaps essential, ingredient for life since it enables complex chemistry and the recycling of carbon dioxide, giving the planet a protective and warming blanket. Astronomers have discovered over a dozen "super-Earths," a designation that ranges up to twice our size and 10 times our mass. They're likely to be raucously active, with crashing plates, bursting hot springs and geysers, and a frenetic carbon cycle. Super-Earths pump the accelerator of life hard, although it's unclear whether or not larger life-forms could handle the action.[6]

Habitability evolves as a planet does. The fate of planets isn't written in stone because stones change. Geological and chemical processes sculpt all but the very tiny planets. Regardless of the star it orbits, each large planet holds an energy source in the radioactive decay of heavy elements deep in its core. Thanks to this power pack, planets have "lives" of their own. If biology starts, it immediately begins to alter its environment. Our Solar System started with three habitable terrestrial planets. Two suffered runaway climate changes that left them barren— one remade itself to stay habitable.

The Man in the Moon

The terrestrial planets are the natural subject of attention in a search for other living worlds but the moons of the gas giants have provided hints of habitability since they were surveyed by the Voyager probes. Although the gas giants lie beyond the region where liquid water can exist on a moon's surface, there are local energy sources—radioactive heating from interior rock and tidal heating from the gas giant.

As of late 2008, astronomers counted 172 moons in the Solar System. Jupiter and Saturn each have about 60, Uranus has 27, Neptune and even the dwarf planets manage 6 among them. The 7 largest stand out: Ganymede, Titan (both larger than Mercury), Callisto, Io, Earth's Moon, Europa, and Triton. Titan orbits Saturn, Triton orbits Neptune has 13, and the other four apart from our Moon are the moons that Galileo discovered orbiting Jupiter 400 years ago.[7] The largest moons rival small planets in size and mass and a couple have active geology and substantial atmospheres.

Europa is an eerie world; its surface is a jigsaw puzzle of crumpled ice sheets under a tenuous oxygen atmosphere. This far from the Sun the ice is as hard as granite. But tidal heating from nearby Jupiter provides enough energy to keep a salty, liquid ocean 100 kilometers deep under the ice. The environment is no more extreme than Antarctica's under-ice Lake Vostok, creating excitement about the possibility of microbial life in Europa's ocean. NASA's budget woes have stalled a mission to land on the ice, melt through it, and study the ocean with a hydrobot. Bob Pappalardo probably speaks for most planetary scientists when he expresses frustration with this situation: "We've spent quite a bit of time and effort trying to understand if Mars was once a habitable environment. But Europa, potentially, has all the ingredients for life . . . and not just four billion years ago . . . but today."[8]

Titan is the other siren of the outer Solar System. The Cassini orbiter and the Huygens lander showed us a familiar landscape of seas and river deltas and ice floes and clouds, but with a bizarre twist. Titan's active weather is based on ethane, methane, ammonia, and acetylene. Volcanoes spew ammonia mixed with water, and hydrocarbon lakes are seen at the northern latitudes. The atmosphere is thicker than Earth's and made almost entirely of nitrogen. Titan is an amazingly promising lab for prebiotic chemistry and, perhaps, a place where we might find life quite unlike familiar forms on Earth.

Apart from these two compelling moons, models by Adam Showman from the Lunar and Planetary Lab suggest 8 to 10 additional moons in the outer Solar System where water might be kept liquid by pressure under a rock and ice crust, so microbial life can't be ruled out. Even tiny Enceladus—500 kilometers across—displayed eruptions of subsurface water when *Cassini* flew by in 2005.[9] This naturally leads to speculation about habitable moons in the vast amount of real estate under investigation by the people who search for extrasolar planets. In *Return of the Jedi*, the Ewoks race through a terrestrial-looking landscape on the Forest Moon of Endor, in pursuit of the minions of Darth Vader, while the planet orbits a gas giant like Jupiter. Perhaps science fiction isn't as outlandish as we think.

The search for exoplanets has netted a large number of Jupiters and super-Jupiters and a smaller number the size of Uranus and Neptune. The same formation process that formed moons around our gas giants should operate in distant solar systems so astronomers are estimating how many distant planets will host habitable moons. There are several requirements. A mass of 7 percent or more of Earth's is needed to hold onto an atmosphere. If the moon is less than 25 percent of Earth's mass it will need tidal heating to be geologically active. They shouldn't have orbits that are too eccentric or "days" so long they cause temperature extremes. Many environments will satisfy these criteria.

Even though earths are just beyond reach for the big planet surveys, there's a prospect of being able to detect Earth-sized moons around gas giants. About 10 percent of the exoplanets found so far orbit within the habitable zone of their host star. As more are found the odds rise that some will transit the star as seen from our perspective. The wobble of the planet caused by the large moon makes the planet vary its position and velocity, a subtle signature that will show up in the timing of the eclipses. This challenging observation is well motivated because moons can be more habitable than planets in the same location.

Counting Habitable Worlds

First, a caveat: Counting habitable worlds in the Milky Way is an abuse of induction. We are still gathering enough information to determine whether terrestrial planets and giant moons in our Solar System are habitable. Extrapolating to the fragmentary information on exoplanets and across the vast Galaxy is audacious bordering on foolhardy. But here goes.

The raw material for any calculation is the 400 billion stars in the Milky Way. That number is dominated by low-mass red dwarfs, perhaps 90 percent of the total, so their ability to shelter life is crucial. In 2005, the SETI Institute held a workshop to specifically address habitability of dwarfs. The standard concerns have been that a planet would have to be very close to such feeble stars to keep warm, and at that distance it would lock tidally to the star and always keep one side to the star. But Earth-sized planets will have thick enough atmospheres to equalize the heat flow. The other issue is a wafer-thin habitable zone compared to a star like the Sun. But there are so many dwarfs that in the aggregate they contribute almost as much habitable real estate as the smaller number of Sun-like stars (Figure 7.4). If planets are massive and heated

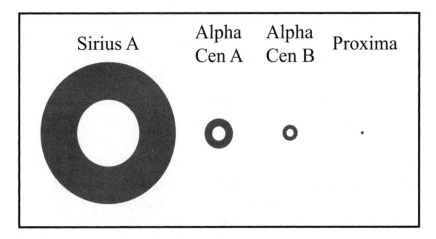

Figure 7.4. Traditional definitions of a habitable zone refer to a shell of space around a star where the inner and outer radii are set by the boiling and freezing points of water, respectively. Sirius A is a bright, luminous star with a large habitable zone; the three stars to the right are all in a triple system very close to our Solar System. The habitable zone of the red dwarf Proxima Centauri is a third the size of the dot on the "i" in Proxima.

from the inside, their distance from the dwarf or any star is irrelevant and numerically dominant dwarf stars will contribute the most habitable planets.

Some researchers have argued for the existence of a galactic habitable zone in addition to the circumstellar habitable zone. It's supposed that regions too near the center of the Galaxy suffer greater threats due to supernovae and stellar interactions, while regions at the periphery are too poor in heavy elements to make planets. But it turns out that stars migrate substantially in radius within the Galaxy over billions of years so the galactic habitable zone can't be a hard constraint.

Traditional calculations toss out the 50 percent of all stars that are in binary systems, but simulations show that most of these are so wide that any terrestrial planets are undisturbed. Then there's the still-unmeasured number of terrestrial planets per distant star system. We're forced to rely on computer simulations of how these systems form, where an average number of two to four planets within a factor of 2 of Earth mass are seen, most of which have a lot of water—between 1/10 and 100 times the contents of our oceans. To add in habitable moons we have to use the Solar System as a guide. The range is two to four, with the low number counting only Europa and Titan and the high number adding the Galilean moons Callisto and Ganymede.

Where does this leave the census? With traditional views of habitable zones, we saw earlier an estimate of 100 million habitable worlds. If we admit most red dwarfs and binaries, but accept the factor of 10 reduction for the galactic habitable zone, the raw material is 20 billion stars living 2 billion years or longer, which is plenty of time for life to emerge. With roughly 6 habitable worlds per system, it's a staggering 100 billion potentially habitable worlds in the Galaxy.

The last two factors are complete guesswork. Suppose just 10 percent of the habitable worlds actually develop life, and on only 1 percent of those does it evolve to become multicellular and complex.[10] The Milky Way then may contain 100 million worlds hosting advanced life of unknown function and form. The eventual demise of life in the Solar System will be a small setback in a galaxy that's permeated with life. A bustling Milky Way will barely miss us.

Threats from Beyond

Trouble in the Neighborhood

Earth orbits the Sun in a kind of shooting gallery, and as we've seen, asteroids and comets take pot shots at us from time to time. Most of the major impacts are from asteroids; a one-kilometer asteroid hits Earth about once every half million years while a similar-sized long-period comet only hits once every 30 million years. Comets are lethal because they travel much faster than asteroids so pound-for-pound they pack more punch. Also they arrive from any direction so are harder to spot than asteroids, which are found close to the ecliptic plane. Last, there's a vast reservoir of comets with the potential to pay a visit.

The reservoir is a tenuous spherical halo called the Oort Cloud. The hypothetical Oort Cloud surrounds the Solar System and extends

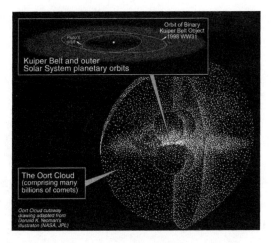

Figure 7.5. The Oort Cloud is a comet reservoir extending to 50,000 or 100,000 times the Earth-Sun distance, which is a significant fraction of the distance to the nearest stars. When stars approach or penetrate the Oort Cloud, the gravitational disturbance can send many additional comets inward on Earth-crossing orbits.

to 1000 times further than the distance of Neptune from the Sun. It's hypothetical because it's inferred from comet orbits (Figure 7.5). By Kepler's second law, comets on highly squashed elliptical orbits travel slowly when they're far from the Sun and then they zip through the inner Solar System. They're active when they're close to the Sun and very faint the rest of the time. The Oort Cloud is suspected to hold a trillion comets, and even though that adds up to only five times Earth's mass, it's a lot of ammo for the shooting gallery.

Rich Muller, a physicist at the University of California at Berkeley, thinks the Sun has an unruly bully for a smaller brother. Back in the early 1980s he saw a hint of periodicity in mass extinctions in the last 500 million years. Random extinctions have many possible causes but extinctions like clockwork beg for an astronomical interpretation. Muller hypothesized a dwarf stellar companion to the Sun moving in a lazy orbit but getting close enough every 26 million years to shake up the comets at the edge of the Solar System, sending some cascading toward us. He called the star Nemesis.

"Give me a million dollars, and I'll find it," he says. Nobody has taken him up on that, but even though he's now retired, he spends much of his spare time looking for the dim red star that he thinks is the Sun's missing companion. Nemesis has been attacked as a *deus ex machina* explanation—it features in a book called *Nine Crazy Ideas in Science: A Few Might Even Be True*, by Robert Erlich, along with "distributing guns reduces crime" and "radiation exposure is good for you." More substantial problems for the theory are the lack of consensus that extinctions are periodic and the fact that there's only strong evidence for an extraterrestrial origin with the K-T event. Even if Nemesis is never found, there are other neighbors that can cause us trouble.

The Oort Cloud is delicate and skittish, and any time it's affected by the gravity of a nearby object, it convulses and sends comets on potentially Earth-crossing orbits. This doesn't happen all at once; it's spread out for 2 to 3 million years after the triggering event. Researchers have used catalogs of nearby star motions from the Hipparcos satellite to predict how often stars will "brush up" against the Oort Cloud. They found an average of four stellar passages within three light-years per million years. The effect on the comet cloud depends on the speed as well as the distance, so the biggest effect is often from the fastest interloper rather than the closest.

Gliese 710 is an unassuming red dwarf in the Ophiuchus constellation. It's faint enough that you'd need binoculars to spot it. But data from *Hipparcos* show it's making a beeline for us at 25 kilometers per second (55,000 miles per hour), arriving in 1.5 million years. At that time it will blaze in the sky as bright as the belt stars in Orion. Gliese 710 is predicted to pass just over one light-year from the Sun and nudge several million comets into orbits that will cross Earth's. It sounds like a lot, but it's only a 50 percent increase in the normal number. Over that same time span eight stars will get closer than our current nearest neighbor, Proxima Centauri. Barnard's Star will be knocking on the door in just 10,000 years.

Stars that actually penetrate the Oort Cloud are expected every ten million years. They could scatter comets like tenpins and cause a real hazard for Earth. While the threat is real, recent research suggests that comet impacts don't come in sharp pulses.[11] This million-comet simulation followed over the age of the Solar System found that the gravity of the Milky Way acts to smooth out influences due to passing stars, leaving the comet influx rate relatively constant.

When Massive Stars Die

The Solar System is part of a larger galactic ecology. We're 26,000 light-years from the center of the Milky Way, traveling in the plane of the disk on a leisurely 220 million year orbit. Although the Sun and its neighbors share circular orbits, nearby stars approach and recede from us, we pass through molecular clouds where new stars form, and in and out of the spiral arms of the Milky Way. We undulate through the plane of the Galaxy every 30 million years, which is coincidentally close to the claimed timescale for periodic extinctions.

After stellar encounters, the biggest threat to life on Earth comes when a massive star dies violently as a supernova. Supernovae are bringers of life and, if you're sitting too close, bringers of death. Thanks to the heavy elements they've ejected into the interstellar medium, new stars have enough silicon, aluminum, and iron to make planets and enough carbon, nitrogen, and oxygen to make life. In fact, a nearby dying star probably triggered the collapse of the solar nebula that led to us.

When a massive star exhausts its nuclear fuels, there's no pressure to support the enormous weight of the gas so the core collapses. It then bounces, creating a blast wave at a temperature of billions of degrees and releasing a torrent of neutrinos, gamma rays, and cosmic rays. An incredible 10^{58} neutrinos are emitted at light speed, and the brightness increases by a factor of billions so the dying star rivals a whole galaxy.

A supernova goes off somewhere in the Milky Way every 50 years but the Galaxy is a big place so the odds of one anywhere near us are very low. Could ancient supernovae have caused extinctions? Astronomers predicted that the gold-plated signature of such an event would be the deposition of radioactive isotopes that could only come from a dying star. Unfortunately, a "smoking star" will always be lacking in this kind of research. After thousands of years, the glowing gas left behind by a supernova fades, and the pulsar that often forms gets such a kick that it could have moved far from the site of the explosion.

In the late 1990s, a group from Germany tested the method with radioactive iron-60 measurements in deep-sea rocks. In 2004, using better data, they found clear evidence for iron-60 enhancement in a layer dating to 2.8 million years ago.[12] This was a time when fossils of marine fauna show a mini-extinction (Figure 7.6). The supernova may have been anywhere from 30 to 300 light-years away. It's fascinating that researchers can use deep-sea ocean sediments as a telescope to see the nuclear fires of long-dead stars!

That's in the past, but what about the future? So many people are trying to scare us (including me) that it's nice when something can be taken off the worry pile. In 2003, refined calculations of the effect of gamma rays on Earth's ozone layer showed that a supernova would have to go off within 25 light-years to destabilize our biosphere, and that

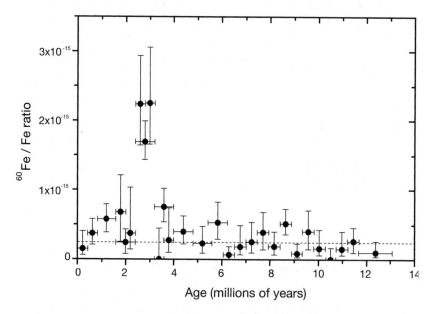

Figure 7.6. A mini-extinction of marine fauna about 2 to 3 million years ago aligned well with a peak in the deposition of a radioactive isotope of iron that is produced in supernovae and then ejected into space. The graph shows the fractional abundance of radioactive iron relative to normal iron plotted against time before the present day in millions of years. No other trace of the death of the star would be visible after so long, but this type of evidence ties cosmic cataclysm to terrestrial catastrophe.

drops the incidence rate to 1 every 700 million years. Because we have a good stellar census within that distance and no star is massive enough to die as a supernova, we can breathe a sigh of relief and drop that particular alert from amber to green.

While we're on the subject of excessive fear factors, black holes make many people slightly nervous. Neil Tyson's book *Death by Black Hole* archly fueled this unease, although only one of his essays was actually about black holes. When a massive star dies as a supernova it leaves behind a collapsed core from which nothing can escape. Should the endless night of a black hole cause us to lose endless nights of sleep?

Probably not. Contrary to the popular conception, black holes aren't cosmic vacuum cleaners, sweeping up everything in their path and sucking in passing strangers. Space and time are only distorted in the immediate vicinity of the event horizon. At any distance beyond a few thousand miles the gravity would be no different from that of a star of the same mass. Given the vastness of space, you'd only be at risk if you willfully ventured close. You might, for example, die by falling into a vat of molten iron, but only if you located the smelting plant in your area, drove there, got past all the security, and climbed on the railing of the catwalk above the vat, having drunk too much beforehand.

Future space travelers should be able to avoid them too. They're rare, because only a small fraction of stars die as supernovae. It would take billions of point-to-point journeys across the Milky Way to have a significant chance of encountering a black hole. If you did, the result would not be pretty. The strong tidal force would stretch you to breaking point. Long before you reached the event horizon, you'd be squeezed like toothpaste from a tube and elongated into many fragments, extruded like biological spaghetti. This would be a joy ride to hell.

Meanwhile, those infuriating physicists are stoking our fears by trying to make black holes with their accelerators, right? Actually, no. It's the popular media who are irrationally enamored by the idea, attracting eyeballs and raising eyebrows with headlines like "Doomsday Machine Destroys World." Under some but not all theories of matter, the Large Hadron Collider could create tiny black holes, but so tiny they would evaporate in seconds. And even if they did escape the collider and fall to the center of Earth, the extra material they accreted would only delay their evaporation by a few more seconds.[13]

The Ultimate Cataclysm

Occasionally, the universe is so surprising that even clever scientists find it hard to imagine how nature could create the phenomenon they observe. How about this: A dying star that releases as much energy as 1000 Suns would over their entire lifetimes. Where for a few seconds it outshines the rest of the universe in gamma rays, and creates a blast wave so bright that these single stars can be seen 13 billion light-years away, from a time not long after the big bang and as far away as the most distant galaxies. That's a gamma ray burster.

They eerily connect a human and a cosmic mode of destruction. In the 1960s, satellites watching for violations of the recent Nuclear Test Ban Treaty saw unexpected gamma ray flashes in the sky. We can imagine tense times while we allayed our suspicions and fears that the Soviets were testing weapons in space. But it turned out that the gamma rays were of cosmic origin. The flashes were so brief and had to accurately locate in the sky that it took 30 years to solve the puzzle, although we don't yet fully understand these bizarre cataclysms.[14]

A gamma ray burst goes off somewhere in the universe about once a day. Modern satellites are able to home in on the gamma rays in a few seconds and direct ground-based telescopes to make a quick follow-up before the fireball fades from view. There are two types. Those lasting less than two seconds are thought to be caused by the collision of two neutron stars. Those lasting longer are believed to be examples of a particularly violent supernova explosion called a hypernova, which is the shriek of a star taking its last breath and the resulting black hole taking its first.

When a rapidly rotating star 20 times the mass of the Sun dies, it undergoes core collapse to form a supernova, leaving behind a black hole, but with extra oomph. The high mass and temperature compared to a normal supernova mean gamma rays pour off the blast wave that results. The rapid rotation means that radiation and hot gas surge out along the rotation axis of the black hole, drilling through the top layers of the dying star and sending twin jets into the universe at 99.995 percent of the speed of light. We only see the gamma ray bursts when one of the jets is pointed at us, which means there are hundreds of times more that we don't see.

The rate of gamma ray bursts is one per galaxy per 100,000 years, so they're thousands of times less frequent than supernovae. But the energy and so the potential threat are correspondingly greater. What would happen if we found ourselves in the beam of one that went off 1,000 light-years away? Fasten your seat belts.

The fireball would shine as brightly as the Sun in the sky. Moments later, the flood of high-energy radiation would hit the atmosphere, setting it ablaze. Forests would burn, lakes and rivers would boil off, and the side of Earth that faced the blast would be sterilized. The shock wave from the impact would send a mile-high wall of flames around the planet, perhaps sparing some ocean organisms on the far side. Intense gamma rays and UV radiation would destroy the ozone layer worldwide. It's unlikely anything could survive this onslaught.

The good news about gamma ray bursts: their ferocity is concentrated in the narrow beams of radiation. Such an event occurring in the Milky Way *and* being pointed in our direction happens once every 100 million years or so. That's intriguingly close to the average time between the big extinctions. In 2003, Adrian Melott from the University of Kansas and his collaborators suggested that a gamma ray burst might have caused the Ordovician extinction 450 million years ago.[15] They didn't have any physical evidence to join the two ideas, so it was more of a plausibility argument than a hypothesis with evidence to support it. The timing argument implies that such an event must have affected Earth several times in its history.

When *Discover* magazine listed the top 20 threats to life on Earth, gamma ray bursts came second, just after asteroids. Phil Plait, well-known for his *Bad Astronomy* blog, has also done a service to humanity by giving a ranked list of threats in his recent book *Death from the Skies!: These Are the Ways the World Will End*. He puts the odds of fatality per lifetime from a gamma ray burst at 1 in 14,000,000—20 times lower than an asteroid impact.

Let's look at the color of their danger level. Stars massive enough to go as a hypernova are very rare, but we need a full stellar census out to 1000 light-years, and our knowledge of the Milky Way isn't detailed enough to be sure none exist that far out. Eta Carina is on the watch list. Just visible to the naked eye, it's 8000 light-years away and the most luminous star in the Galaxy, putting out as much energy in five seconds as the Sun does in a year. We've seen it double in brightness just 10 years ago, so it's unstable, and it's 100 times the Sun's mass, so it's likely to die as a gamma ray burst. Luckily, its spin axis doesn't seem to be pointed at us.

Another star called WR 104 is more ominous. It's a similar distance away but aimed right at us.[16] "I can't help a twinge of feeling that it's uncannily like looking down a rifle barrel," says Peter Tuthill, who was the first researcher to identify the threat. "We probably have hundreds of thousands of years before it blows, plenty of time to come up with some answers." And that will have to do as far as reassurance goes; radiation from a gamma ray burst gives no advance warning of its arrival, so there's really no point in worrying.

For endings, it's hard to beat quick, painless, and high in drama. If humans are taken out by a gamma ray burst, it would be nature's biggest good-bye, visible across the universe. We can't really take it personally if we lie in the path of such a cataclysm. Much as it wasn't personal in Douglas Adams's *A Hitchhiker's Guide to the Galaxy* when Earth found itself in the way of a galactic construction project.

Chapter 8

THE SUN'S DEMISE

Sallie Baliunas is the Sun doctor. She knows when the Sun sneezes, the Earth gets sick. She measures the light pulse and magnetic "breathing" of the Sun and studies the often subtle, sometimes profound, effects on our planet. Back in the seventeenth century, the Earth caught a bad cold. This period was called the "Little Ice Age," when temperatures in Europe got so low that the Dutch skated on canals all summer and Scots were snowed in all year. It was also a time of solar inactivity, when the Sun's magnetic field was so scrambled that it "forgot" its sunspot cycle.

Baliunas caught the astronomy bug in 1977, when she was a grad student working at the Mount Wilson Observatory in southern California. On her first night observing, a lightning bolt shattered a nearby tree and blew out all the windows in the building; she took that as an omen. She's one of a dedicated cadre of professional astronomers and volunteers who saved the 2.5-meter (100-inch) telescope—a leviathan of its time that Edwin

Hubble used to measure the size and expansion of the universe—from imminent closure. This grand old telescope is once again used every clear night for research and Baliunas is the site manager.

Watching the Sun and waiting for it to age is worse than watching paint dry, so Baliunas studies Sun-like stars at different stages of their evolution to piece together a story of how the Sun will behave as it ages. This gives her a chance of using a large sample to track changes that take much longer than a lifetime.

Baliunas has entered the public eye in an uncomfortable way for all people who like to think of scientists as dispassionate, objective, and above the political fray. She thinks most climate variations are caused by the Sun rather than by human activity; she's argued against the Kyoto Protocol and the conclusions of the Intergovernmental Panel on Climate Change. She's the darling of conservative think tanks and has had research funding from ExxonMobil and the American Petroleum Institute.

As tempting as it would be to dismiss her as a corporate shill who is swimming against the overwhelming current of evidence on climate change, Baliunas embodies some interesting questions about the process of science: How often are scientists influenced by political ideology when they interpret data? Isn't it healthy to have prevailing explanations subject to skepticism? What if there are multiple mechanisms involved in climate change? And by not learning about all the effects on our planet of our star—which powers the biosphere and sustains all life on Earth—what if we're throwing the baby out with the bath water?

Living with a Star

Stormy Weather

In *Gulliver's Travels*, Jonathan Swift told of the floating island of Laputa—a mythical land of philosophers and astronomers who were obsessed with the sky. They fretted that the face of the Sun would increasingly be covered by spots to the point where it wouldn't give sufficient heat and light to the world. When Laputians met someone early in the day their first question wasn't, how are you? but, how did the *Sun* look this morning?

Two thousand years ago, Chinese court astronomers noted and kept track of blemishes on the face of the Sun. For all cultures, where the prediction of weather was based on folklore and seasonal variations, any clear connection between the appearance or behavior of the Sun and climate was of great practical importance. In 1801, Sir William Herschel reported his discovery of a relationship between the number of sunspots and the price of a bushel of wheat. When sunspots were scarce the price of wheat was always higher. He reasoned that fewer spots on the Sun meant a deficit in the radiation emitted, leading to poorer growing conditions, diminished agricultural production, and, through the inexorable law of supply and demand, higher prices.[1]

Two hundred years later, the Sun still puzzles scientists as they try to understand its effects on Earth. The total energy arriving at the top of the atmosphere is 1366 watts per square meter; think of a lightbulb shining through each area the size of a piece of paper. Over the 11-year solar cycle, and as long as we've been measuring it accurately, solar output varies by only 0.1 percent. That's rock steady; you'd be hard-pressed to notice the variation in a lightbulb that changed from 99.9 watts to 100.1 watts. Naively this would only change temperature by 0.05°C (0.09°F), but climate models are very uncertain and different versions predict the Sun could cause between 10 and 30 percent of the recent global warming.

Modest light variations conceal a deeper truth: the Sun is a dynamo—in both senses of the word. Sunspots are areas of reduced surface temperature that mark intense concentration of magnetic fields. The entire magnetic field of the Sun flips each peak in the sunspot cycle. Even though it's a small fraction of the total solar output, the short wavelength radiation varies dramatically, with complex but profound effects on Earth's atmosphere and climate. UVB radiation varies 15 times as much as visible light over a solar cycle, and these energetic photons have a substantial impact on the ozone layer.

Secondary effects associated with sunspots are prodigious: loops of plasma larger than Earth thrown off the surface, reconnecting magnetic field lines that release the energy of a billion atomic bombs, and huge outpourings of gas from the surface, called coronal mass ejections. When flares occur, astronauts on the space station must huddle in a shielded area to avoid cell damage. When coronal mass ejections happen, we have several days to prepare for the assault on our satellites and power systems. Mild ejections pump up the auroras to make a pretty light show; the most severe ejections cause billions of dollars of damage to the power grid.

Unless you're out in space, none of this behavior is life-threatening, but the violence is a reminder that the Sun is anything but a dull and steady star. A once-in-a-millenium flare in 1859 set off an aurora so bright that people in England read by its light and so extensive that it was seen in the Bahamas and Hawaii. Some Sun-like stars undergo "superflares," with energies 10 to 100 million times the strongest flare observed on the Sun; they would almost certainly destroy the ozone layer and so disrupt the food chain. Luckily for us they only seem to happen when a giant planet in a close orbit gets its magnetic field tangled with the star, and puny Mercury has no magnetic field.

NASA has bundled its solar programs under the title "Living with a Star" (as if we really had any choice). The flagship mission is SOHO (*Solar and Heliospheric Observatory*). This venerable satellite has transformed our view of the Sun's interior, its exterior, and the solar wind. And in its "spare" time it's discovered over 1000 Sun-grazing comets. The Global Oscillation Network Group runs a set of six solar imagers around the globe that continuously monitor the Sun, which "rings" like a bell. Unraveling the harmonics gives a detailed map of the interior.

Solar modelers use powerful computers to crunch grids of data into a three-dimensional view of the Sun. They can even predict the coronal mass ejections, when magnetic field lines get so twisted they snap like a rubber band, flinging billions of tons of plasma toward Earth at a million miles per hour. They were spot-on in their prediction of the appearance of the Sun's corona during the March 2006 total solar eclipse.

The Sun's amazing power and its centrality to our existence registers most vividly during a solar eclipse. While not an eclipse "junkie" like some of my colleagues, I've been lucky enough to see several total eclipses. They're particularly impressive when they occur on water. From near the bridge on a cruise ship off Baja, California, in 1991, I watched the shadow streak across open water toward us, sending hundreds of spectators into an awed silence. At noon in the tropics, it was like a fist had punched a black hole in the sky overhead. In 2006, I was off the coast of Turkey as we steamed for a hole in the clouds to catch the eclipse. We were a hundred miles from the site of a famous eclipse in 585 BC, when Herodotus wrote that two warring tribes, the Medes and the Lydians, became terrified and confused when the sky darkened. They put down their weapons in the middle of battle and declared peace. I watched in amusement as the spectators crowded the railing and jostled each other for a better view, although one side of the ship wasn't really any nearer the darkened Sun than the other.

The Subtlety of Fire and Ice

Londoners skating on the Thames. Swiss glaciers advancing on farms and crushing entire villages. The Swedish Army marching across 32 kilometers of frozen sea to invade Denmark. The Dutch fleet trapped in a harbor for six months. Iceland's population dropping by half as kilometers of sea ice embraced the island, strangling the economy. New Yorkers walking across the ice from Manhattan to Staten Island. Birds dropping from tree branches, frozen to death.

These were scenes from a "Little Ice Age" that caused severe winters across northern latitudes in the middle of the seventeenth century. It coincides roughly with something called the Maunder minimum, the period from 1645 to 1715 when sunspot numbers were unusually low. You might think that the Sun without acne would make it warmer on Earth because sunspots are dark, but the bright regions that surround them will more than compensate so the Sun is actually dimmer in a sunspot minimum. However, the full story is more complicated, because a pristine Sun will only cool temperatures by 0.5°C (1°F), less than experienced in Europe. Moreover, the Little Ice Age was confined to the Northern Hemisphere and it began as early as the thirteenth century, with distinct temperature minima in 1650, 1770, and 1850. Increased volcanic activity played a part, and possibly a shutdown of the "great ocean conveyor" that lets Europe enjoy the warming of the Gulf Stream.

Over longer timescales, changes in the Sun's output were substantial but they had to be measured indirectly before the age of telescopes. Radioactive carbon-14 is an excellent proxy. It's created when cosmic rays from space bombard oxygen in the upper atmosphere, then it is incorporated into tree rings. Another proxy is beryllium-10, which can be measured in polar ice layers. The solar magnetic field shields the Earth from cosmic rays, so when solar activity and sunspots are low more carbon-14 and beryllium-10 are created and deposited and the concentration goes down when solar activity is high. The two proxies confirm Maunder's minimum and show that solar activity over the past 70 years is as high as at any time for 8000 years. The implications for climate aren't clear because the coupling is indirect and complex.

Ten thousand years is a drop in the bucket of geological time and the past ten millennia have been a warm period called the Holocene that followed the peak of the last Ice Age 20,000 years ago. Pristine ice and sediment layers give us an unbroken record of climate changes over 5 million years (Figure 8.1). The last million years in particular have been a time of dramatic climate variations, with swings of over 10°C (18°F)—larger than the amount we're worrying about with the current, human-induced global warming.

The Sun is partly responsible for these variations, in a subtle way

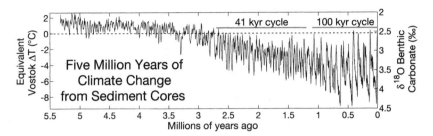

Figure 8.1. Global climate variations over the past 5 million years, as measured from 57 sites of deep-sea sediment cores, sampling the radioactive isotope concentration in tiny cold-water protists (the right scale, which plots the concentration of a temperature-sensitive mineral metabolized by organisms in the oceans). This can be related to global temperature variations sampled from Vostok, Antarctica, ice cores (the left scale, which projects temperature change globally from samples of the ice sheet in one location). Regions where the 41,000 and 100,000 year Milanković cycles appear are marked.

that relates to changes in Earth's orbit. A hundred years ago, Serbian mathematician Milutin Milanković speculated that ice ages were caused by a set of cyclic variations: the 100,000-year variation in the degree by which Earth's orbit of the Sun departs from a circle, the 41,000-year change in the tilt of Earth's axis relative to the Sun, and the 23,000-year meander of Earth's spin axis as the planet wobbles like a top. All of these effects change the amount of radiation received by a particular place on the surface. Cosmic rhythms combine to make a complex pattern of heating and cooling.

As with all climate variations, other factors like volcanism, impacts, and ocean circulation must play a role, because ice ages aren't as regular as Milanković's theory predicts, and the 100,000-year cycle has been the strongest over the past million years whereas the theory says it should be the weakest.[2] Regardless of the mechanism, our ancestors during the past few million years wouldn't have complained about a little global warming.

A Really Big Finale

These hiccups and eructations are just a way to dance around a larger truth: the Sun won't live forever. For the first time since we looked at how we die, we're faced with the inevitable rather than just the probable. The Sun's demise will create a huge headache for us, or any creatures that might still be around when the time comes.

Calling the Sun ordinary is true in a simplistic sense; there are stars more and less massive, hotter and cooler, longer and shorter lived. Ranked by mass or brightness, it's quite impressive, in the top 15 percent. And there's nothing ordinary about a ball of gas that a million Earths could fit into with room left over, that transforms 635 million metric tons of hydrogen into helium each second, that has converted 100 earths of mass into pure energy by Einstein's nifty equation $E = mc^2$, and that powers our biosphere even though we're 150 million kilometers away and intercept less than a billionth of its radiation.

The Sun gets its energy from the same fusion process that we use in our most fearsome weapons. Yet the Sun isn't a bomb; it stays puffed up and a constant size because everywhere within it there's a perfect balance between gravity pulling inward and the pressure from fusion reactions pushing outward. As the fusion product helium accumulates in the Sun's core, it's compressed and heated up, increasing the overall fusion rate. So the Sun will steadily get larger and brighter, as it has for 4.5 billion years (Figure 8.2).

Jim Kasting at Penn State has run the best models of how the Sun will brighten as it ages. In half a billion years, warming will accelerate the weathering rate, moving carbon dioxide from the atmosphere into the oceans. Ironically—given current concern over CO_2-induced warming—too little CO_2 is even worse news. The concentration of

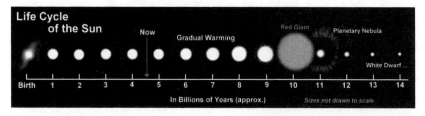

Figure 8.2. The life cycle of the Sun, shown schematically. We are roughly midway through the main sequence phase of hydrogen fusing into helium. Thereafter, the Sun goes through a series of disruptive changes as it adjusts its configuration and finds new energy sources. Eventually, all nuclear fuels are exhausted and the Sun cools forever as a white dwarf.

carbon dioxide will drop so low that photosynthesis in most trees and plants will stop. Some plants that concentrate the gas for themselves, like sugar cane and tropical grasses, will soldier on, but the biosphere will be mortally wounded.

After that it gets fairly calamitous. The ice caps will melt and equatorial regions will flood and become too hot for many animals. Water from the very warm oceans will rise into the stratosphere and start leaking into space. The oceans will steadily evaporate. The entire planet will become waterless desert. Relentless radiation from the warming Sun can then unlock carbon dioxide from bare sediments on the ocean floor, warming the planet even more. Finally, even that gas is baked off into space. About 3.5 billion years from now, a visitor would find Earth to be a desiccated rock.

The chronology has a strange and terrible symmetry to it, placing us at the apex and center of the history of life. As Peter Ward and Don Brownlee noted in *The Life and Death of Planet Earth*, the microbial dominance that gave rise to complex life will return as animals and plants succumb to the brutal conditions. If species can evolve to be optimized for living underground or under the seabed, the biosphere may persist into a time when the surface makes the Atacama Desert look like Eden. The bookend to the last common ancestor will be the last common descendant, perhaps a tough survivor like *Desulforudis audaxviator*, deep in rock, bathed in a radioactive glow.

The Sun isn't massive enough to die as a supernova, but it has a few tricks left. The core will be squeezed and heated by unfused hydrogen until 5.5 billion years from now when that shell of hydrogen ignites and the Sun becomes a subgiant about twice its present size. Then, about 700 million years later, it's the finale. Not the actual end, but the climactic passage with sizzling violins, crashing cymbals, and the brass section giving it all they've got. The Sun's core compresses and heats up until it can't be compressed any more. The heat drives out the cooler layers of the Sun, turning it into a bloated red giant. At its peak, it will be 250 times bigger and 2700 times more luminous than it is today.

What happens to Earth depends on a race between two effects. As its outer layers balloon out, the Sun sloughs off a third of its mass, so its gravitational grip weakens and the planets start to spiral outward. But the outer layers, which are a toasty 2760°C (5000°F), are streaking out through the Solar System toward Earth. The outcome will be decided in a scant 7.6 billion years, so don't forget to make plans for your estate.

Recent calculations paint a bleak picture. As seen from Earth, the blood-red Sun will grow until it looms large and then fills the sky. Even though we're receding, the Sun is approaching us faster. For a thrilling time we will skim the flame tops. Then friction from travel through the Sun's outer layers will slow our trajectory and we will inexorably spiral inward. Earth goes down.

Goodbye, Crazy Diamond

The coda to this symphony seems moot. Earth, Venus, Mercury are engulfed and reeled in by friction and gravity. Mars escapes by the skin of its teeth—the Sun has consumed its closest children. But let's press on because it's still our story; the vaporized Earth is part of the Sun's heavy heart.

What happens when the end comes? Unlike people, stars don't slow down as they age. They expire in Byzantine, spasmodic rhapsody. Our guide to the death of the Sun is Caty Pilachowski. She was a senior staff member at the National Observatories for 20 years and she knows as much about the inner lives of stars as anyone. This slender woman, barely five feet tall, her head framed by a halo of dark curls, has been a great role model for women in astronomy. As president of the astronomer's professional society, she sat on two phone books to get a better view of her colleagues, and ran the meetings with calm, firm authority.

She describes Earth's outcome equably. "We'll end up in the Sun, vaporizing and blending our material with that of the Sun. Part of the Sun then blows away into space, so one might say Earth is cremated and the ashes are scattered into interstellar space." As at every stage of a star's life, gravity and pressure due to nuclear reactions stand in opposition. But unlike the standoff during 11 billion years of turning hydrogen into helium, now temporary advantage is claimed by one side and then the other. Arm wrestlers had sweaty fists locked overhead but now their arms veer dangerously from one side to the other.

The Sun is schizophrenic. Its outer envelope cools and recedes into space while its core shrinks, reaching the insane temperature of 100,000,000°C (180,000,000°F). To stave off gravity, it fires up a new thermonuclear reaction, fusing helium into carbon. It marks this new transition with a terrific flash where it momentarily exceeds its normal energy output by a factor of 100 billion. Then it shrinks to a dimmer, smaller, and hotter state than it experienced as a red giant. Creating helium plus a little bit of oxygen and neon will buy it a hundred million more years of life. The core is now 300,000,000°C (540,000,000°F).

When the helium runs out, gravity gains the upper hand and forces a new core collapse, triggering violent and unstable fusion. A series of four convulsions occurs, each separated by just 100,000 years. Fusion reactions ripple through the outer layers of the Sun and it reaches the peak luminosity of its life, 5200 times its current brightness. These are the dying gasps. The paroxysms of the expiring Sun will briefly light up filaments of ejected gas in the outer Solar System and beyond; the red and green glow will be a modest version of the beautiful planetary nebulae imaged by the *Hubble Space Telescope*. It's not clear what the best vantage point for this show would be. Mars has been engulfed. Europa's oceans have been vaporized and Titan has been charred. Maybe those who were smart enough to buy real estate on Neptune's moon Triton could watch the denouement in relative comfort.

Shorn of its envelope and denied any new nuclear fuel, the Sun's core contracts to a few thousand miles across. It's brilliantly bright and hot, over 200,000°C (360,000°F)—a white dwarf. When Pink Floyd penned "Shine On, You Crazy Diamond," it was a double metaphor. It was an homage to Syd Barrett, a founding member of the group who fizzed brightly and then dissipated in a murk of drugs and insanity. It was also a paean to the late stages of stellar evolution.

The carbon-rich material of this cooling ember exists in a strange state halfway between graphite vapor and diamond, so dense that a handful would weigh as much as a jumbo jet. Pyrotechnics behind it, the Sun leaches its remaining heat into space and gracefully fades from white to yellow to dull red and then to black.[3]

Moving Off-Earth

Our Future in Space

If Earth and Sun finally fail us, and some future species is smart enough to do something about it, the best bet will be off-world. Space travel up to this point has been fairly primitive. Rockets are glorified fireworks and the first astronauts had to wear diapers. About 500 people have experienced the thrill of slipping the bonds of Earth, and 5 percent of them have died (to keep it in perspective, that's less than the 10 percent death rate among roughly 2700 people who've climbed Everest). For many futurists and visionaries, the space program has been disappointing, the few high peaks outnumbered by valleys and long, flat plains.

Speaking personally, I've always been happy to do without my food in a pill and my Dick Tracy TV watch and my hover car. But I do *so* want to experience space or set foot on another world before I die. With the glacial pace of space exploration, and my inability to fork over $20 million to the Russian Space Agency, it looks like those pleasures will wait for my children or, more likely, my grandchildren.

What went wrong? The space program was born out of superpower rivalry and for decades was operated exclusively by the government agencies of just two countries. That's not a recipe for efficiency or innovation. As a way of learning about the universe, remote sensing gives us information much more cheaply than space exploration. Any large telescope can see stars across 95 percent of cosmic time, map out the architecture of galaxies, measure the abundance of elements trillions of times rarer than hydrogen, and detect planets around distant stars. By contrast, NASA limps along with the 30-year-old space shuttle, where two out of five have been lost to catastrophic failures, with the death of 14 crew members. With spacecraft, they've managed to put 12 men on the Moon and haul back 1000 pounds of rocks, but it was so hard and expensive we've not been back for nearly 40 years.

That's unfair, but not completely. For its first three decades the space program was a proxy competition in the Cold War. Now we're seeing a shadow of that in the ascendancy of China as a space power, but the larger landscape is the involvement of private and commercial players. Space exploration is leaving its painful birth phase and prospects for the future are very bright.

In the United States, the door was opened by the 1984 Commercial Space Launch Act and a 1990 law that deregulated space and made a level playing field between NASA and commercial operators. By 1997, Russia had privatized most of its launch capability as well. NASA has recognized that it needs private partners to realize its goals; in 2006 the agency announced $500 million of financing for development by the private sector and in 2008 it awarded the initial contracts. Early culture clashes between entrepreneurs and government civil servants are already generating some heat!

The incubator of the private space program was the Mojave Desert, where Burt Rutan was dreaming of space. Like his hero Wernher von Braun, he started building rockets as a kid. His designs transformed the way light aircraft are made and smashed long-distance records for small powered planes. Rutan founded Scaled Composites in 1982 and in 2004 the experimental aircraft he built in the desert won the Ansari X Prize for the first pair of flights to near-space by a reusable manned spacecraft. Rutan wasn't in it for the money; it cost over $100 million to win the $10 million X Prize, but the idea to spur competition with a very public prize was modeled after successful aviation prizes of the early twentieth century.

The floodgates are open. Internet entrepreneurs are piling in to invest in space, and sometimes just paying to be space tourists. The FAA has registered 18 companies to work on low-cost launchers. The goal of orbital flight has been achieved, and a company called Space Island Group is planning to build a private space station with rented living quarters. Payload launch costs have dropped from $10,000 per kilogram to $1000, or about $70,000 for an adult, and if they drop much further the dream of a ride into space will be within reach of many people, including me and you. The latest competition is the Google Lunar X Prize, nicknamed Moon 2.0, which will bestow $20 million on any team that can land and operate a lunar rover by the end of 2012. Progress in the past decade has been breathtaking.

The new space visionaries aren't wild-eyed dreamers; they're sober scientists and engineers like my college buddy Robert Bond. He used to work for the U.K. Atomic Energy lab at Culham on fusion as an energy source, but he got fed up with government bureaucracy and the fitful progress toward the goal, so he left his civil service job and now works for Reaction Engines Limited, a private company designing orbital and suborbital vehicles that use a hybrid air-breathing ramjet and liquid-fuel rocket. This company and others have had to navigate the politics of European Union countries that are heavily invested in the traditional nonreusable rockets.

On a mild summer day, we drink beers in his garden near Oxford and he's animated as he describes the technology that he hopes will make space travel routine. Robert is a gentle man who plays rhythm guitar in a 1970s rock cover band on the weekends, but space isn't a hobby for him, it's a quest. We tilt our heads back and look into the pale blue of an English summer and imagine the future.

It no longer seems unreasonable to claim that our destiny is in space. Stephen Hawking thinks so and he has a seat booked on *SpaceShipTwo*, a joint venture between Rutan's company and Richard Branson's Virgin Galactic. Hawking says, "I don't think the human race will survive the next thousand years, unless we spread into space. There are too many accidents that can befall life on a single planet. But I'm an optimist. We will reach out to the stars."[4] Despite the cost and the practical difficulties, space still has the power to inspire. Recall the words of Antoine de Saint-Exupery: "If you want to build a ship, don't drum up people together to collect wood and don't assign them tasks and work, but rather teach them to long for the endless immensity of the sea."

What about bypassing all this heavy lifting and "beaming" people off the planet to another location, a ruse used by science fiction for 50 years? Teleportation is the transmission of someone's full information, atom by atom, to a remote location at the speed of light. Having this technology would also enable suspended animation and backup copies in case something went wrong with the original. For a long time it was thought that the entanglement of quantum states would prevent any transmission of quantum bits of information, or qubits. But in 1993, IBM researcher Charles Bennett published an elegant demonstration that teleportation was possible in principle.[5]

In a lab, it's not quite that easy. In 2009, Christopher Monroe and his colleagues at the Joint Quantum Institute cooled two ytterbium atoms to within a fraction of a degree of absolute zero and used microwaves to put them into an entangled state.[6] He then used photons to "read" the state of each atom, although they were separated by a meter. The result showed that quantum information could actually be transmitted. But it's not *Star Trek* yet. Only 1 of every 100 million teleportation attempts succeeded and it took 10 minutes to send a bit of quantum information one meter. "We need to work on that," said Monroe.

Brave New Worlds

At least we know what we're facing. On a timescale of a billion years, we'll see the merging of the continents (again), a drop in the level of carbon dioxide to below the level needed to support life, followed by the oceans boiling, the surface baking and being sterilized, and then the eventual death spiral and disintegration as our planet falls into the Sun. Astronomers have seen the future and it's not pretty. In 2008, a white dwarf called GD 362 was found to have rocky debris orbiting it, the probable remains of shredded planets.

What if we could move to a better home? Terraforming is an idea that has blended smoothly from science fiction to science proximity. Olaf Stapledon provided the first fictional description in 1930, when Venus was terraformed in his pioneering *First and Last Men*. More recently, terraforming is at the heart of Kim Stanley Robinson's Mars trilogy. The scientific discussion of terraforming was pioneered by Carl Sagan, who wrote about alterations to Venus in 1961 and Mars in 1973.

It's defined as the process of modifying a planet or moon environment to be inhabitable by humans.

Given that the Sun is going to get warmer, let's take Venus off the table and focus on Mars. Mars is 50 percent farther from the Sun so moving there would buy us a lot of time, or provide a refuge if we completely mess up Earth. Reengineering a planet is a formidable undertaking but NASA takes the idea seriously enough to have hosted conferences on it and funded early design studies (Figure 8.3).

There are three stages to terraforming. First, a planet has to be made amenable for life. The surface of Mars is cold, arid, and believed to be sterile, so this means raising the temperature and creating a thicker atmosphere. Robert Zubrin, the founder of the Mars Society, and Chris McKay, an astrobiologist at NASA's Ames Research Center, have worked out how it might be done in detail. The cheapest method will likely cost several hundred billion dollars and take 50 years, all just

Figure 8.3. A Mars base would be the first step toward terraforming the red planet to make it suitable for human habitation. The cost of setting up even a simple base would be hundreds of billions of dollars and the Earth-to-Mars supply line is also extremely expensive, so all terraforming plans depend on mining local materials and doing construction on Mars to gradually make the environment habitable.

to create a basic level of microbial habitability. That's enough to make you blink, but remember, we're trying to save the world![7]

Mirrors equivalent to a single mirror 80 kilometers wide could be built on Mars and positioned over its south pole to vaporize frozen carbon dioxide, but the engineering involved is challenging. Asteroids containing ammonia and water could be steered toward Mars from the outer Solar System, cleverly using the frozen volatile gases as fuel for rockets that would deliver the asteroid. Comets might be steered into Mars in the same way. Both types of impact would have a useful side effect of releasing nitrogen from Martian soil to act as a buffer in the new atmosphere. The last method involves power plants on the surface, manufacturing chlorofluorocarbons to act as potent greenhouse gases. Those gases could also be compressed and delivered by rocket from Earth.

In terraforming Mars the aim is to use positive feedback in our favor. As carbon dioxide is released from its frozen storage in the poles, it acts as a greenhouse gas, warming the planet and accelerating the melting of the pole. Zubrin and McKay think the atmosphere can be thickened and the surface heated enough to establish standing water and a hydrological cycle. At this point, aquifers might be tapped with little more than pumps and drilling rigs.

The next phase would be to establish a biosphere, most probably using hardy extremophiles that have been genetically engineered for the job. They would be radiation-resistant and oxygen-producing, to start the third stage: the task of making an atmosphere breathable by humans. However, at the second stage people may be able to live on Mars with breathing equipment but without pressure suits, and large inflatable living areas could be built. Rendering Mars fully Earth-like would take thousands of years or, if we had to get it done quicker, we could just spend more money.

The visionaries who make schemes for terraforming are level-headed scientists and engineers, but even they get giddy at times. Creating a new home for billions of humans will be vastly expensive, and getting there will almost certainly require space elevators for the first and last leg to the surface at either end and an armada of highly efficient rockets to act as a "bus" service in between. So it's worth thinking of other strategies.[8]

Engineering the Future

We've already discussed the odds that a passing star could jostle the comet cloud enough to send many of them in our direction. But on the much longer timescales of the Sun's future evolution, there's a chance that a nearby star will pass through the Solar System close enough to eject Earth entirely. Given what we'll face, that might not be a bad outcome, but there's only a 1 in 100,000 chance of it happening before the Sun becomes a red giant. There's an even smaller chance, 1 in 2 million, that we'll be captured by a passing star and so gain a new home and a new lease on life. As they say, hope isn't a strategy.

What about moving Earth to a safer location deliberately? That's a tall order, far beyond what we could achieve by getting the population of China to jump at once (which has no effect at all) or all point firecrackers overhead at the same time. We would have to corral a huge asteroid and send it toward Earth—not hitting us but coming close enough to act as a gravitational slingshot and nudge us outward. Do we really trust rocket scientists enough to pull off a stunt like that?

It's not as far-fetched as you might imagine. NASA has cut its teeth on the method already, using Jupiter and Saturn to "pump up" the Galileo and Cassini probes and send them into the deep Solar System. Here's how it works. If we send a probe on a trajectory catching up with the planet from behind, it will gain some energy from the planet's orbital motion. So the probe speeds up and the planet slows down, moving a bit closer to the Sun. Jupiter is so massive it never misses the energy, but it helps propel *Galileo* on its way. It works the other way too. Send a probe ahead of the planet and the probe loses energy and gives it up to the planet, which moves slightly further from the Sun.

Don Korycansky, Greg Laughlin, and Fred Adams are scientists who've worked out the details and published them, straight-faced, in the peer-reviewed astronomy literature.[9] All it takes is an asteroid the size of Long Island with (solar-powered) rockets attached to steer it within 16,000 kilometers, and Earth would be nudged 16 kilometers outward from the Sun. That's not much, so we have to reuse the asteroid, which means sending it past Jupiter or Saturn to gain back some energy and then swinging in by Earth for another pass. It may take 10 years for each pass; after a million passes we'd have moved Earth out to the distance of Mars. Essentially, we're turning the asteroid into a shuttle conveying energy from Jupiter or Saturn and delivering it to Earth to increase the size of its orbit.

Cute, but *very* risky. At a distance of 16,000 kilometers, the asteroid would be terrifying, looming almost as large as the Moon in the sky. Its tidal force would be 10 times that of the Moon, causing tsunamis and big storms. Each time it zipped by we'd have to batten down the hatches. We would have to do it a million times and each time the margin for error would be small. A miscalculation leading to a collision would, the authors note dryly, "sterilize the biosphere most effectively, at least to the level of bacteria." As Laughlin points out, "There are profound ethical issues involved, and the cost of failure is unacceptably high." Let's hope we don't get to the point where it's a choice we have to make.

The Transhumans Are Coming

What type of people will get to live out these futuristic fantasies of the Solar System eons in the future? We've already encountered those who are putting themselves on ice in the hope of being resuscitated using a technology not yet invented, and Ray Kurzweil, who foresees a time not too far off when we will transcend our biological bodies. These are just threads in a larger philosophical movement called transhumanism.

Transhumanism, often abbreviated H+, is an international movement that explores the use of science and technology to enhance our mental and physical capabilities and overcome aspects of the human condition such as disease, aging, and involuntary death. It's a hypothetical "you" as if you'd been merged with your fantasy sports car (Figure 8.4).

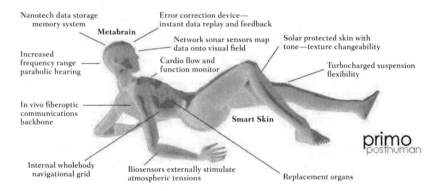

Figure 8.4. A hypothetical posthuman, showing enhancements all designed to promote longevity or even immortality. Many of these are advanced versions of technologies we are beginning to develop now.

Although the movement is unabashedly futurist, it echoes the ancient themes of the Gilgamesh story, and it has unbroken threads to some of the great thinkers of Renaissance humanism. In the modern era it was influenced by computer scientist Marvin Minsky and it coalesced around a set of academics at UCLA in the 1980s. One of them was a man called FM-2030 (formerly F. M. Esfandiary), a Persian who wrote fiction and nonfiction and who's one of the people frozen by the Alcor Life Extension Foundation. He once said, "I am a 21st century person who was accidentally launched in the 20th. I have a deep nostalgia about the future."

Transhumanism has also influenced feminism through the writing of Donna Haraway. Her *Cyborg Manifesto* moved away from Oedipal or Christian narratives to embrace a humanism that transcends gender duality.[10] The World Transhumanist Association was founded in 1988, with 5500 members in 100 countries. It recently changed its name to Humanity+. Its newsletter of the same name is edited by R. U. Sirius (formerly Ken Goffman), who ran for the presidency of the United States in 2000 for the Revolution party.

Transhumanism is controversial and often misunderstood. The main thread of transhumanist thought focuses on the risks as well as the benefits of altering humans to live better and longer lives, and it also pays attention to potential inequality of access to these technolo-

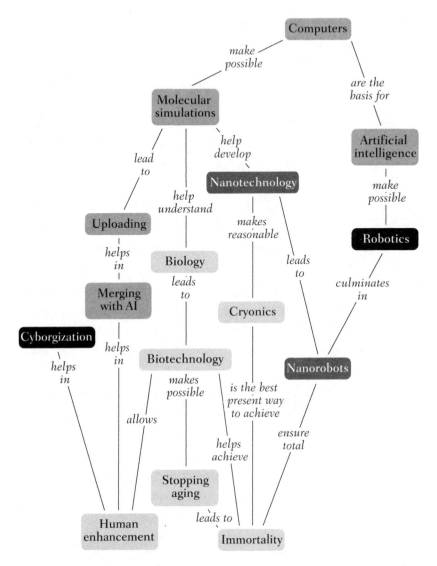

Figure 8.5. Domain model for transhumanism. This shows the paths by which research on new technologies might address the underlying issues of aging and biological death. The incremental strategy calls for enhancement; the ultimate goal is immortality.

gies. Geneticist J. B. S. Haldane noted back in 1932 that every advance in biology or genetics would "first appear to someone as blasphemy or perversion." There are not many movements that could be described by a critic, Francis Fukuyama, as "the world's most dangerous idea," and by a supporter, Ronald Bailey, as the "movement that epitomizes the most daring, courageous, imaginative, and idealistic aspirations of humanity." A conceptual map of the subject shows the technologies it will depend on to be realized (Figure 8.5).

Coming from Nick Bostrom, these ideas sound sober and sensible and it seems almost immature not to consider them seriously. Bostrom is cofounder of the World Transhumanist Association, a professor of philosophy, and director of the Future of Humanity Institute at Oxford University. He believes that any death prior to the heat death of the universe is premature. He thinks that technology to "upload" minds independent of the biological brain is a real possibility. "You can freeze the brain and slice it up then scan each slice with some microscopic technique, and then use image processing software to extract a 3D map of the neuronal network that your brain runs." A 140-paged paper on his Web site outlines a roadmap to whole-brain emulation.[11] Bostrom thinks that most academics give short shrift to such topics and his goal is to "attempt to make it possible for humanity, rationally and thoughtfully, to consider its own future and approach these challenges with greater wisdom."

Its wilder practitioners do sound very strange, but transhumanism faces issues that will be front and center in the twenty-first century. Its main declaration says that by "embracing new technology we have a better chance of turning it to our advantage than if we try to ban or prohibit it." The movement isn't anthropocentric; it advocates "the well-being of all sentience, whether in artificial intellects, humans, posthumans, or non-human animals." A more interesting question than, *what is the future?* may be, *who is the future?*

Chapter 9

OUR GALACTIC HABITAT

The Milky Way is human emotion writ large in the sky. As the ancient Greeks told it, Helios had an impetuous son, Phaeton, who was sure he was strong enough to drive the chariot of the Sun across the sky. Helios tried to convince him otherwise but Phaeton grabbed the reins and commandeered the fiery chariot and its four horses. The horses could tell an unsure hand was guiding them so they bolted and lurched through the heavens, leaving Earth alternately frozen and scorched as they veered high and low.

The people in India and Africa were so badly burned that their skin stayed dark. Zeus contained the damage by hurling a thunderbolt at the boy, who fell to Earth as a meteor and landed in the Euphrates River. The gods mourned his death by placing the river up in the sky where it became the constellation Eridanus. The irregular scar burned across the sky by the chariot became the Milky Way.

Other legends tie the diffuse band of celestial light to the pangs of jeal-

ousy. One of the earliest Greek stories concerns Cronus, the father of Zeus. Not wanting to lose his position as sky god, Cronus swallowed his children. Rhea, the Earth, could not bear losing another infant to her husband's jealousy so she wrapped a stone in swaddling and gave it to Cronus to swallow. Suspicious, he asked her to nurse the child one more time before he devoured it. Rhea pressed the hard rock against her nipple and the spurting milk became the Milky Way.

Another version of the story has the infant Hercules suckling milk from Hera to gain her wisdom. When she realized the child was the bastard son of Zeus and another woman, she pushed the baby away and a smear of milk turned into the Milky Way. This story is immortalized in a luminous painting by the Venetian master Tintoretto in London's National Gallery.

In other parts of the world there were traditions to venerate the Milky Way that transcended individual cultures. People in East Asia believed that the hazy band of stars was the "silvery river" to heaven. In the Vietnamese version, the Weaver Fairy weaves silk robes and the Buffalo Boy tends the herds. They fall in love but neglect their duties so are punished by the Jade Emperor to live on opposite sides of the silver river. They are the stars Vega and Altair. The Jade Emperor relents and allows them to meet once a year, on the seventh day of the seventh month, the auspicious festival of Qi Xi.

The Milky Way featured prominently in Mesoamerican cosmogonies. The Maya called it the World Tree, where the star clouds of the Galaxy are the tree of life. At dawn in mid-August, the Milky Way stands erect, running through the zenith, the axis of the raised-up sky. As the sky turns and *seven stars of the Big Dipper set, in Mayan myth the Seven Macaws are knocked off their perch atop the tree of life.*

Once and Future Stars

The Mayfly and the Forest

The mayfly is born in water and emerges for its day in the Sun—quite literally. After a year as a water nymph, it will spread its wings for a time measured in hours. Imagine that one particular mayfly is a freak of nature, endowed with the ability to fly through the forest, take in what it sees, and cogitate about the implications. What would it see, and what could it learn about the forest in such a short time?

It would see other insects and animals but they would be dwarfed by the complex, verdant landscape. On the forest floor, mosses and flowers nestle in undergrowth and bushy vegetation. Saplings rise higher and some merge into the canopy formed by towering trees. A few trees are bare, their bark fallen away. Others are charred or split. The mayfly would also spot fallen logs. With no leaves and branches, they might not be recognizable as segments of the sturdy trees that rise into the sky. Some of these logs might be so degraded by beetles and weathering that they're more similar to dirt than to trees.

Perhaps the mayfly would conclude the forest was really all about the insects and animals. After all, apart from a few leaves falling and the flowers that might turn to follow the Sun, the plants and trees aren't doing much. Meanwhile the animals are foraging and digging burrows, the bees are flitting from flower to flower, and ants are industriously ferrying food and building materials along complex paths on the leafy forest floor. This purposeful activity seems to take place against a static backdrop. To a squirrel, an oak tree is nothing more than a home and a source of food in the form of acorns.

Could the mayfly deduce from observations during its short life that many of the items in its world are temporally related? That an acorn can grow into a sapling, that the tiny saplings are precursors to trees 30 meters tall and 9 meters in girth, that the trees die and fall to become logs, and that the logs over time are degraded into the mulch that lies under the carpet of leaves? Could the mayfly figure out the pattern of germination of planets and pollination of flowers? Could the mayfly be aware that the acorn is food and a home for the larvae of weevils and so partakes in another part of the cycle of forest life? That caterpillars or beetles can fell a mighty oak just as surely as lightning?

With such a short life span it might take luck. If the mayfly were born during a thunderstorm, it might learn that some trees get struck by lightning and die. If the mayfly happened to see an acorn fall it might deduce that the acorn had some purpose after it fell, but it probably couldn't guess that only 1 in 10,000 acorns grows into an oak tree. And even if a tree fell once a day somewhere in the forest, it's unlikely to be close enough for the mayfly to witness it.

Such is the problem of short-lived astronomers in a long-lived galaxy. Mayfly lives are millions of times shorter than the lives of oak trees and our lives are millions (or even billions) of times shorter than the lives of the stars. The Galaxy presents an almost unchanging tableau.[1] We can't see rotation or evolution and each type of celestial object seems distinct and unique. It takes careful observations and physical models of how its stellar denizens work to understand the Milky Way in which we are embedded.

A City of Stars

The dim band of light that girdles Earth has been known since antiquity. The Milky Way—translated from the Latin *Via Lactea*, which derives from the Greek *Galaxias* and its etymology in milk—bisects the celestial sphere, running as far north as the constellation of Cassiopeia and as far south as the constellation of Crux. The brightest region lies below the equator in the direction of Sagittarius. Its light isn't smooth or regular; the band flares and shrinks as it traverses the sky and it's

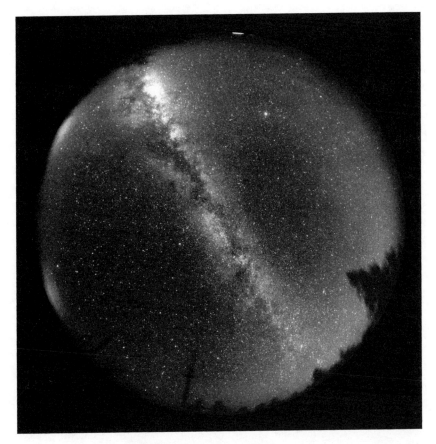

Figure 9.1. The Milky Way arching overhead, as seen from a remote site in southern Arizona. People who live in urban or suburban areas are never treated to a sky dark enough to view the Milky Way, and a view like this is getting increasingly rare due to light pollution.

superposed with knots of bright stars and ragged dark regions (Figure 9.1). This was all common knowledge to the ancients, but is mostly forgotten now that most of us live in places awash with the glow of civilization. Few people will forget the first time they saw a truly dark sky and the Milky Way leapt out at them like a braided white curtain.

What is the Milky Way? Democritus guessed that the gauzy glow was the combined light of distant stars (although there's evidence he had been influenced in this idea by Anaxagoras). He also speculated on the existence of invisibly small, fundamental units of matter that he called atoms. Democritus was reputed to have laughed so readily at almost everything that he was sent to Hippocrates to be cured. Hip-

pocrates said that Democritus wasn't mad; he just had a happy disposition. Perhaps anyone who could hold so many stars and atoms in his
head would smile. Democritus was brilliant and far ahead of his time
but he had no evidence to support his speculations.

The main current of Greek thought discarded this idea in favor of a
geocentric cosmology, where Earth was immobile within a set of crystalline spheres, with the interior spheres carrying the Moon, Sun, and
five naked-eye planets, and the outer sphere carrying all the stars. This
system was initially proposed by Eudoxus, but Aristotle was its most
eloquent and forceful proponent. Aristotle found it self-evident that
Earth didn't move, that the celestial objects moved in perfect circles,
and that we were the center of everything. The stars formed an amphi-
theater with us on center stage. We should rearrange the letters slightly
and call it the egocentric cosmology.

In geocentric cosmology, the stars must be at an equal distance
from Earth to explain the fact that they don't vary in brightness or position over the course of a year. In the most complex geocentric model, 56
crystalline spheres were required to account for the sometimes subtle
motions of the planets; they were offset and nested like the gears of a
heavenly watch. In the version of the model passed on by Ptolemy in
his masterwork *The Almagest*, the distance to the outermost sphere was
about a million miles, so it was whirling around at a prodigious clip, 110
kilometers per second (250,000 miles per second).

That's just the bright stars; what about the band of light we call the
Milky Way? Aristotle was pretty far off the mark. He thought that the
nebulosity of the Milky Way was produced by ignition of the uppermost
layer of air by the same mechanism that produces the tail of a comet.[2]

Nearly 2000 years later, Galileo began experimenting with the
newly invented telescope. He revealed many wonders of the sky—the
spots on the Sun, the moons of Jupiter, and the mountainous terrain
on the Moon—but perhaps his biggest surprise came when he pointed
his telescope at the Milky Way. The gauzy nebulosity sharpened into
myriad points of light just as a smooth TV image turns into individual
tiny phosphor dots or CCD pixels when you get very close. To Galileo,
this gave the sky real depth, a third dimension. It seemed reasonable

to suppose that each dot of light was a star like the Sun and that some were much further away than others. Galileo saw stars too numerous to count and presumed that there were many more beyond the grasp of his small telescope. We live in a city of stars.

I remember well when I first saw—*really* saw—the Milky Way. I'd been an astronomer for several years and had observed on mountain-tops in North America. Those sites were dark but there were always cities on the horizon, bleeding their light upward into the night sky. As a postdoc, I went to Chile for the first time, and the flight north from Santiago and the rugged five-hour drive on dirt roads told me I would be far from civilization. The observatory was on a promontory of rock and all around was wilderness: high Andes peaks, arid brown foothills like a crumpled blanket, and the southern edge of the Atacama Desert. But nothing prepared me for the sky when I went out at midnight. The Milky Way arched overhead like a frayed silver rope. Its star clusters and dark clouds had depth and texture. The starlight blazed so bright that I could read the book in my hand and they both cast a shadow onto the ground. I stood still for several minutes, stunned.

Architecture of the Galaxy

Galileo couldn't measure the size of the city. That was left to William Herschel, who took telescope building to new heights near the end of the eighteenth century. Herschel was a professional musician and self-taught astronomer who gained fame and royal patronage by his discovery of Uranus. Working with his devoted sister Caroline, an accomplished astronomer in her own right, he conducted nightly sweeps of the sky and counted stars. The density of stars declined as he moved away from the band of light in either direction and was roughly constant along that band, so he inferred that we live near the center of an enormous disk of stars. He estimated the Milky Way to be 8000 light-years across and 1500 light-years thick, containing 300 million stars.

How did he make this estimate? We're trapped on the surface of our planet and can't reach out to measure the third dimension. Herschel made use of the way light travels. Rays from any spherical source of light spread out and diffuse as they travel through space. Seen from afar, the brightness of any light source falls off as the inverse square of the distance. Move two times further away from a star and it will appear four times fainter. By assuming that all stars are intrinsically the same, Herschel used the relative brightness of stars to tell their relative distances.

Herschel was right about the disk, but wrong about the size and our position within it. He seriously underestimated the size of our galaxy by assuming that all stars are like the Sun. In fact, the stars that are easiest to see are the intrinsically bright stars, those more massive than the Sun. Because they can be seen from very large distances, they're much farther away than Sun-like stars of the same brightness would be. Herschel's method of estimating distances was too crude to give him a reliable result.

Now flash forward a century. American astronomer Harlow Shapley took up the challenge of measuring the size of the Milky Way, armed with a new 1.5-meter (60-inch) telescope on Mount Wilson in California, then the world's largest, and a new method of measuring distances using the properties of pulsating stars. Shapley studied tight concentrations of stars called globular clusters and found them to be at amazingly large distances, from 50,000 to 200,000 light-years away. They formed a spherical swarm at the periphery of the Milky Way. The swarm wasn't centered on the Sun but on a position 27,000 light-years away in the direction of Sagittarius.

Shapley was right about the direction of the center of the Milky Way but he also made a mistake about the size. His error was to assume that the space between stars was empty. Space is permeated by a thin gruel of gas and dust. Light scatters off the tiny dust particles and the cumulative effect is that distant stars are dimmed so astronomers are tricked into thinking they're further away than they really are. It took the development of radio astronomy in the 1930s to probe the Galaxy with long waves that are unaffected by interstellar dust.

In the past 30 years, astronomers have refined their estimates of the Milky Way's size, mass, and stellar content. Mass is the biggest surprise because it's 10 times larger than we would expect given the number of stars. The visible contents of the Galaxy are contained and retained by gravity in a halo of dark matter. Dark matter is a great enigma of modern astronomy; its fundamental physical nature is still mysterious yet it keeps all galaxies intact and its particles outnumber all the atoms in the universe.[3]

The Milky Way is roughly 100,000 light-years across and 5000 light-years thick. We say roughly because there's no sharp edge; the stars thin out steadily until they run out in the inky vastness of intergalactic space. The spherical halo is dotted with white dwarfs and the globular clusters but it's mostly made of dark matter. Within it, the disk of the Galaxy is etched with beautiful spiral arms, trailing from the direction of rotation. The central regions contain a pileup of old and red stars and there's a black hole a few million times the mass of the Sun that lurks at the exact center.[4] Seen on its edge, the Milky Way would look like two fried eggs set back to back. Astronomers are fond of homey analogies. Sure, the universe is unsettlingly large, but don't be unnerved, breakfast is served!

Our galaxy contains about 400 billion stars. This number is even more uncertain than the size because it depends how deep you look. Most stars are far smaller and dimmer than the Sun, and they're far more numerous than stars like the Sun.

What about our place in this city of stars, 60 times more numerous than people on the planet? We're within the urban area but far from the center, as Pasadena is to Los Angeles or Wembley is to the City of London. We're close to the Orion spiral arm, a major thoroughfare of stars, and the stories of star birth and death surround us. The center of the Galaxy is a chaotic place, so crowded with stars that if we lived there the night sky would be lit as bright as the full Moon. If we lived near the edge it would be very lonely, offset by the privilege of being able to gaze at the magnificence of the spiral arms.

Birth and Childhood

The Milky Way is vast and it seems eternal. But it had a beginning and it will have an ending. Like the players in Shakespeare's soliloquy from

As You Like It, stars will have their exits and their entrances, and even the stage will not last forever. With the infant, "mewling and puking," we have a connection to the myths and legends of antiquity. The Milky Way hasn't always been the same size. Unlike a baby our galaxy didn't begin as a miniature version of its present self. It was assembled from smaller pieces. It's easy to make a baby—a moment's passion, a slip of the birth-control device. But how easy is it to make a galaxy?

Let's meet Carlos Frenk. Carlos is the Ogden Professor of Physics and director of the Institute for Computational Cosmology at the University of Durham in England. He's a leading practitioner of using computers to simulate aspects of the universe and the evolution of galaxies.

Frenk waves his arms as he describes the work of his institute. He's not just animated, he's exuberant. He has dark, flashing eyes and a strong nose, dark hair, slightly graying, and the smooth and husky accent of his native Argentina. We imagine him as an aging Lothario, yet the objects of his attention are not women, but galaxies whose secrets he craves. We can also imagine the cultural disconnect when he encounters taciturn inhabitants of this small market town in the northeast of England, their arched eyebrows as he buttonholes them at the checkout of the supermarket to explain the beauties of dark matter.

Simulating galaxy formation and evolution is extremely difficult. It may not be actual sorcery but it's just as much an art as a science. Think of it as baking. Into a computer we place ingredients: normal matter, dark matter, radiation, and laws of gravity and gas dynamics. Many complex interactions and reactions will take place.[5] When we come back after a few billion years, will we find a fallen, blackened mess, or a beautiful spiral galaxy frosted with stars?

We can't push a cooking analogy too far. Computational astrophysics is technical and esoteric. What does it mean to simulate a chunk of the universe in a computer where gravity operates so that matter clumps and congeals? There is of course no physical space; it's all done with algorithms and hundreds of thousands of lines of computer code. Even the mathematical space is constantly changing to represent the growth and cooling of the universe since the big bang. And there is no actual matter; the best simulation only has a few billion abstract, computational "particles," which isn't even one per star for a tiny galaxy. Time is compressed and distorted. It takes a few weeks for a powerful cluster of PCs to render the 13.7-billion-year history of the universe but most of that processing time goes into the last few billion years when the universe is large and the number of calculations grows proportionately. Gravity is represented precisely but the complex interactions between matter and radiation must be approximated.[6]

All of this should be enough to make astrophysicists throw up their hands in despair. But when Carlos Frenk throws up his hands at a conference, he's jubilant. "I can't *believe* it works out so well," he exults. "The properties of galaxies all turn out just right—the spiral fraction, stellar populations, angular momentum—it's brilliant!" Not everyone in the room is convinced; they're catching a whiff of snake oil. But there's no doubt that we now know the basic story of how all galaxies, the Milky Way included, got the way they are today.

The Milky Way began to be assembled about a billion years after the big bang. This followed an early phase when the universe was small and dense and too hot for structure to form. There were no stars then so astronomers call it the Dark Ages. Gravity exerted its inexorable grip on the dark matter, sculpting it into invisible clouds that pulled normal atoms into their cores (Figure 9.2).

Gas fell to the centers of these puny concentrations of dark matter, thousands of times less massive than the Milky Way. Much of it collapsed and formed the first stars in the history of the universe. But after a hundred million years or so these stars died, and the massive ones died violently enough to drive out the remaining gas and stop the star formation. There were probably no witnesses to the first light of the universe—the earliest stars were made of pure hydrogen and helium with no heavy elements to make planets or living creatures.

After the infant phase comes childhood, when galaxies grow close to their eventual sizes. The child's "shining morning face" is an era when star formation is vigorous and galaxies are as bright as they'll ever be. The wattage of the universe now is a hundred times below this peak.

Steadily and silently, the dark-matter clouds coalesce into larger and larger units. Within them, crashing gas heats up and forms stars.

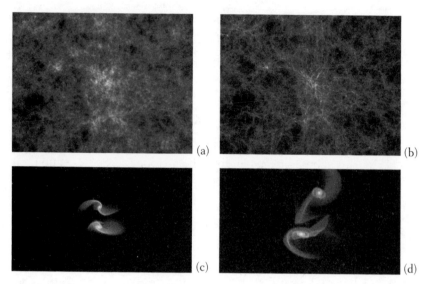

Figure 9.2. Extracts from the Millennium Simulation, the largest-ever computer simulation of large-scale structure, containing 10 billion dark-matter particles and consuming 340,000 hours of CPU time. This view spans 3 million light-years, which is a small region, only 1/1000 of the size of the visible universe. *a*, 300 million years after the big bang the universe is still mostly smooth. *b*, 1 billion years after the big bang dark matter begins to clump into under- and overdense regions. *c*, 5 billion years after the big bang, about 9 billion years ago, many galaxies form where the dark matter has concentrated. *d*, In the present day universe, stars, galaxies, and clusters are abundant.

The Milky Way probably formed from hundreds of tinier dark-matter clouds and as smaller clouds are mopped up, the process accelerates. Final assembly of the Milky Way probably involved some major mergers of moderate-sized galaxies.[7] The overall system had a slight rotation so gas within it naturally collapsed along the rotation axis into a disk. For a few billion years the Milky Way glowed with incandescent brightness as it feverishly formed stars.

Why didn't this process continue until all matter in the universe was mopped up into a handful of supergalaxies? In part because as gas got consumed there was less available for new star formation, and in part because the universe was expanding so galaxies combined less frequently. Galaxy formation is self-limiting so there are few galaxies much bigger than the Milky Way.

Carlos Frenk is off to the side as all this happens, or rather outside, tending the computer where it has been created. He's happy; the simulated Milky Way looks exactly like the real one. "Yes," he says, "amazing. Yes!"

Exits and Entrances

By the time the universe is a third of its current age the Milky Way is no longer an ingénue. She's mature and sophisticated, a stage fully worthy of the pageant of life and death she hosts. Naturally there is heat and light because this is the age of the lover; the Bard tells us she "sighs like a furnace."

We think of stars as lightbulbs but that's not quite right. Deep within every star is a furnace that's combining atomic nuclei under conditions of fantastic pressure. Fusion in the Sun converts hydrogen into helium. Fusion in stars more massive than the Sun converts helium to carbon and carbon to magnesium and silicon, and fusion in the most massive stars converts magnesium and silicon to iron. All of the possible ways that atomic nuclei can combine act to populate the periodic table, and that's how we have the calcium for our bones, the neon for our bright lights, and the copper for our pennies.

The alchemy of a star was best encapsulated by Albert Einstein. His equation $E = mc^2$ describes how a tiny amount of mass is equal to a huge amount of energy. When hydrogen is turned into helium in the core of the Sun 0.7 percent of the mass of the hydrogen atoms turns into radiant energy.[8] This tiny fraction eventually leaks out into space as starlight, much as light might leak out through a factory's windows. Stars aren't just lightbulbs; they're chemical factories. Their most important job is the transmutation of elements.

Lovers can be selfish and if stars were selfish I couldn't have written this and you couldn't read it. The leitmotif of every star is a titanic tussle between the inward force of gravity and the outward force of radiation pressure caused by nuclear reactions. In the Sun, as you read this, it's a standoff—the Sun is neither shrinking nor expanding and it will maintain its current size for billions of years. But when the nuclear fuel is exhausted the Sun and all stars must find a new stable configuration. Their attempts to do this throw off gas into space and that gas is enriched with all the heavy elements the stars had created in their cores. The most massive stars end their lives as supernovae where most of the enriched gas is flung into space.

In a forest, we can witness the circle of life. Stars also take part in a cycle. Each star is a chemical factory and each star will lose some of its material into interstellar space. That gas may occupy the depths of space for a very long time, but in most urban or suburban regions of a galaxy there's enough gas that it will eventually collapse to form new stars. Those stars then build on the inventory of elements from all the preceding generations of stars. Over time, the concentration of heavy elements in any galaxy steadily increases. If stars had instead been selfish, all of the carbon ever created would be locked in stellar cores and biology would be hypothetical.

The Milky Way stage has seen many entrances and exits in the 12 billion years since it formed. Most stars are born and die with little fanfare because they're modest in size and hundreds of times less luminous than the Sun. They all die as slowly cooling embers called white dwarfs. In fact, stars less than a third of the Sun's mass take longer than 12 billion years to convert all their hydrogen to helium, which means that none of them have ever died in the history of the Milky Way. These dwarf stars are hermits. They exempt themselves from the main action on stage and lead quiet, solitary lives.

Massive stars are prima donnas, coming onstage under the glare of the spotlights, in a prominent region of star formation like the Orion Nebula. Their time on stage is brief. A star 10 times the mass of the Sun lives a thousand times less long than the Sun, and a star 20 times the mass of the Sun barely lives a million years, less time than we've been human. Massive stars are generous, promiscuous if you prefer. They shed outer layers late in their lives and fling out most of the rest of their mass at the end in a supernova explosion, leaving only the darkness of a neutron star or a black hole (Figure 9.3).

What does this have to do with us? Every carbon, nitrogen, and

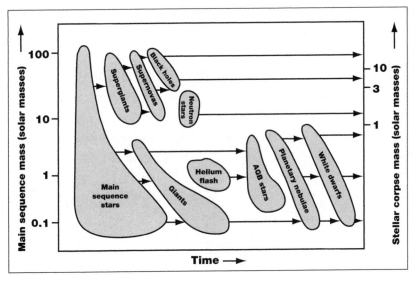

Figure 9.3. Schematic view of stellar evolution, with time running left to right, and mass increasing vertically. There are many more stars of low mass than high mass, and high-mass stars have the shortest lives. Massive stars die explosively, leaving behind neutron stars or black holes, and low-mass stars have violent late phases but die quietly as white dwarfs.

oxygen atom in your DNA—the backbone of all Earthly life—was once inside a star far off in space. Those atoms were spat or sloughed off and spent interminable eons drifting in vacuum, waiting for something interesting to happen. Eventually, they found themselves inside a collapsing cloud of gas and dust, with a ringside seat for the birth of a new star. They were swept into one of a few rocky masses where in a seeming near-miracle they were incorporated into the spindly double helix of life.

We can never know their full story. Atoms are colorless, odorless, and have no taste. They're too simple to bear any imprint of their journey. Some of yours and mine could have been created soon after the big bang and been in and out of the maw of many stars. To quote poet and author Diane Ackerman, we're all "bastards of matter."

In our tale of the Milky Way, Sol makes its entry 8 billion years after the formation. Shame to say it, but there's nothing special about our middle-weight, middle-sized star. It's just one of many players and in 6.5 billion years it will have a final, over-the-top speech and exit stage right. The Milky Way is now an adult. It hasn't changed much since the Solar System formed. Stars in the disk complete a circuit every quarter billion years. Stars in the halo loop in and out of the disk on elliptical orbits. The only action comes from stellar cataclysm. Every fifty years or so a supernova goes off and every 10,000 years a massive star implodes and releases a torrent of gamma rays. Pulsars and black holes sweep their beacons of radiation across the skies. Shakespeare seemed to know; he said the Galaxy is "full of strange oaths."

Mergers and Acquisitions

Dancing with Andromeda

What lies next for our galaxy? Marriage to M31, according to recent calculations. The next time you're out on a dark night in autumn, check out the Andromeda galaxy, M31. First find the great square of Pegasus. The top left star in the square is Alpheratz. Count two bright stars to the left and then two faint stars up. M31 is just above the second faint star and to the right. It's easy to see with binoculars and looks like a sausage-shaped cloud. With the naked eye it's harder and you may need to use averted vision—stare at a nearby star, which brings the more light-sensitive part of your retina into play.

M31 might not look very impressive but it's the most remote object you'll ever see with your naked eye, at a distance of 2.2 million light-years. This barely visible smudge is comfortably remote now. But the gap's closing fast; M31 is approaching us at a blistering 130 kilometers per second (300,000 miles per hour). In 3 billion years we'll merge with our nearest galactic neighbor.

John Dubinski has explored the likely consequences of this merger. A soft-spoken Canadian with long hair and a goatee, Dubinski uses the same techniques as Carlos Frenk to model the cosmic collision. This type of simulation requires enormous computational muscle. Dubinski used all 1152 processors of a machine called Blue Horizon at the San

Diego Supercomputer Center to follow 150 million stars in each galaxy. Even people jaded by the ever-growing power of personal computers may be impressed by the feat—his simulation used a thousand times the memory of a typical PC, created 3,000 DVDs of data daily, and blazed along at a trillion operations per second. Imagine everyone on Earth punching out calculations on each of a hundred calculators every second.

For two billion years, not much happens. The Milky Way gobbles up a few smaller gas-rich neighbors, using the gas to make new stars and quickly assimilating the stars into the disk. But the attraction is undeniable; the two galaxies cannot resist it.

A galaxy collision is entirely different from a collision between people or everyday objects. When the gas components of the two disks meet the gas gets compressed and heated, glowing with the pink light of its dominant element, hydrogen. But when the two sets of stars meet the result is complex and surprising. It's complex because gravity has an infinite reach so each star is affected by all the other stars in both galaxies. It's surprising because the space between stars is so large—think of grains of sand separated in space from each other by several hundred feet.

The two galaxies will pass through each other like ghosts in the night and engage in a complex gravitational dance. This courtship ritual will take several hundred million years. If there's anyone here to witness it, M31 will approach close enough to fill the entire night sky. As we make our first passage through M31, gravity will induce loops and ripples in the distribution of stars and fling off a great arc containing about a billion stars into the intergalactic void.

Another 500 million years will pass and the galaxies will continue to interact and convulse but the bulk of the stars will gradually settle into the center of the combined gravity pit (Figure 9.4). Gas in each galaxy will turned into new stars and the massive ones will die in a Chinese firecracker flurry of supernovae. Both disks will be disrupted and final distribution of stars will have the smooth shape and tight central concentration of an elliptical galaxy, surrounded by the tidal debris of the violent mating ritual. In this strange marriage, two partners will

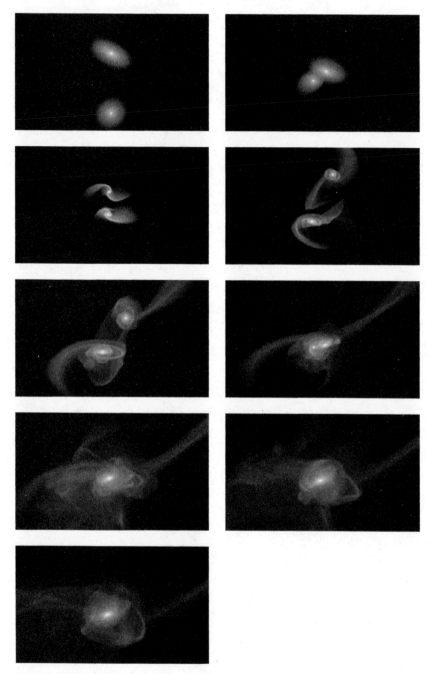

Figure 9.4. Frames from a simulation of the interaction and merger of two massive galaxies like the Milky Way and Andromeda. John Dubinski used the fastest computer in Canada (39th fastest in the world) to see the delicate gravitational interplay of the galaxies, each of which has 150 million "particles." The sequence spans 2.5 billion years.

merge and both will be transformed into a new creature. The courtship, from sidelong glance to consummation, will take over a billion years. There's no point in rushing into things.

Even when one galaxy consumes another, there are subtle traces of the precursors in the orbital motions and chemical abundance patterns of its stars. The Milky Way has devoured a number of dwarf galaxies over the past few billion years and some of their stars are strung out like spaghetti in the halo of our galaxy. Streaks of spaghetti "sauce" tell us what the galaxy has eaten. Astronomers of the very far future should be able to deduce that the single galaxy they see was formed from two spiral galaxies.

John Dubinski has enough of the poet in his soul to realize the beauty spawned by his algorithms. He formed a collaboration with electronic composer John Farah to set the simulated collision to music. Music has been used to successfully evoke the grandeur of space in movies since Stanley Kubrick's *2001: A Space Odyssey* and in countless planetarium shows. Farah and Dubinski released music for a number of realizations of galaxy dynamics in 2006 on a DVD titled *Gravitas*.

The fate of the Sun and Earth in this scenario is difficult to predict because simulations can't reliably track the outcome for a single star. But researchers T. J. Cox and Abi Loeb at the Harvard-Smithsonian Center for Astrophysics have done the math carefully enough to predict odds.[9] During the first enc ounter, a sideswipe 2 billion years from now, there is a 12 percent chance the Sun will take a ride on a tidal tail and be ejected into the depths of intergalactic space, and a 3 percent chance that we'll jump ship and join Andromeda until the merger. The most likely outcome for the Solar System is to be kicked out into the halo, giving us excellent views of the new merged galaxy, which Cox and Loeb dub Milkomeda.

However, it's almost equally likely that we'll be flung toward the pileup of stars in the center of the new galaxy, what the Bard would call its "fair round belly." To follow the next part of the story, we turn to the heart of darkness that lies in every major system of stars.

Dark Heart of the Galaxy

In 1930, Karl Jansky was a radio engineer working at Bell Labs, trying to track down a mysterious source of static that plagued transatlantic phone calls. He built a radio receiver and after filtering out near and distant thunderstorms he was left with a source of radio noise in the sky that rose four minutes earlier each day, which meant it had an astronomical origin. This intense radio emission was coming from the constellation Sagittarius, the direction of the center of the Milky Way.

Astronomers paid little attention to Jansky's work. It was published in a radio engineer's journal and radio technology was so new that most astronomers wouldn't have known how to interpret the results. In the 1960s astronomers also found intense infrared radiation coming from Sagittarius; these long waves can easily penetrate dust and see right to the heart of the Galaxy. When X-ray emission was also discovered, it became clear that stars couldn't be responsible for this concentrated activity. The likely culprit was a supermassive black hole. While a black hole traps all matter and radiation within its event horizon, the intense gravity near the event horizon accelerates matter and causes intense emission at many wavelengths. Theorists speculated that the Milky Way could harbor a black hole far more massive than any star.

On top of a dormant volcano in Hawaii, Andrea Ghez is attempting to measure the black hole with greater precision than ever before. She got her PhD at Caltech and is a full professor at UCLA. Elected to the National Academy of Sciences before she turned 40, and recently named a MacArthur "genius" Fellow, she's emerged as a superstar in her field. Six times a year, she uses the Keck 10-meter telescope in Hawaii to make ultrasharp infrared images of the galactic center. The observatory has a special laser that creates an artificial star in the sky, allowing astronomers to compensate for the blurring effect of turbulent motions in Earth's atmosphere.

But on this particular night, the hardware isn't working and Ghez is nursing a headache. She can't tell if it's because of the altitude—a nose bleed–inducing 4200 meters—or the malfunctioning instrument. Normally, astronomers who use the Keck telescope guide it remotely from the office in Waimea, a pleasant ranch town just 15 minutes from an excellent surfing beach. But when the instrument is new and untested, the hard work gets done on the mountaintop. Ghez grades student papers while engineers swarm over the two-ton instrument. She hopes they'll be gathering photons before dawn.

Ghez wasn't the first to use star motions to estimate the mass of the black hole at the center of our galaxy; she's engaged in a fierce but friendly competition with Reinhard Genzel and his group in Germany. She hopes Keck will give her the edge because its instrumentation allows actual orbits of the stars to be measured as they warp around the black hole at 1300 kilometers per second (3 million miles per hour). A straightforward application of Newton's laws leads to a mass estimate of 4.3 million times the Sun's mass.[10] Since the mass causing the motion fits within the tightest star orbits, only four light-days across, it must be a black hole (Figure 9.5).

In a subject dominated by competitive and occasionally abrasive men, Ghez is very soft-spoken and gracious. She raises a family and spends much more time teaching and working with undergraduates than most researchers of her caliber. She seems to know how lucky she is to be able to think about black holes for a living.

A Bright New Future

After the Milky Way and Andromeda join forces and transform into an elliptical galaxy, something very interesting is expected to take place in the nucleus of the new galaxy. M31 has about a trillion stars so it's more than twice the mass of our galaxy, and it has a black hole 30 times larger or 140 million times the mass of the Sun. In the past 10 years, astronomers have discovered that all massive galaxies harbor black holes. But in most cases, as with the Milky Way, the black hole has a tiny fraction of the mass of the galaxy and there's little gas in the center to eat. As a result, the black hole is dim. If you looked at the Milky Way from afar

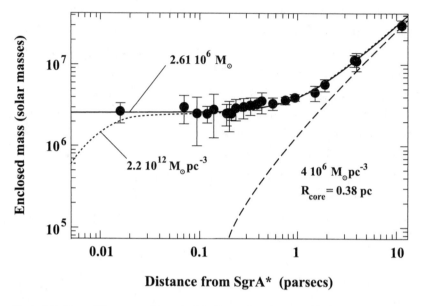

Figure 9.5. Star orbits are used to calculate the enclosed mass with decreasing distance toward the center of the Galaxy, marked by the object called Sagittarius A*. A dashed line shows the continuation of the contribution of a normal star cluster toward the center; there is far more mass than can be explained by gravity of normal stars. The enclosed mass levels off at the projected black hole mass, which has recently been adjusted upward to 4.3 million solar masses.

you'd never suspect anything was amiss in the center of the Galaxy. When a big black hole chows down, it turns into a quasar.

Phil Hopkins is just 30 but he's been building quasars for five years already. He got his PhD at Harvard University and as a young graduate student became the "go to" guy for astronomers who wanted to understand their data. His professor was the head of the Astronomy Department, so Phil learned how to be the sorcerer's apprentice. Phil throws gas, stars, and dark matter into a computer; adds a modest-sized "seed" black hole to the center; and sits back to see what happens. He can speed up time or crank up the black hole mass with a flick of a few lines of code. If a particular calculation proves a bit dull, he can spice it up by throwing in a few galaxy collisions. It's good clean nerdy fun.

The merger that make Milkomeda will send gas clouds crashing into each other, which will trigger them into collapse. Legions of stars will form in their cores. This plays out across the central 5000 light-years of the beefy new galaxy—a burst of star formation that causes those regions to glow intensely. It's a pleasing fire to warm your hands by.

Unseen within this star forming region, because dust mixed with the gas shrouds any outside view, gas near the center will be conflicted. It will be tempted by gravity to fall into a nearby gas cloud, but will feel stronger gravity from the Dark Lord at the epicenter. Gas will fall into the clutches of the black hole so fast that it will grow to a monstrous size, going from a few million times the mass of the Sun to a few billion times in only 100 million years.

Phil chuckles as he described what will happen next. The black hole will have eaten well, and ebulliently it will use its newfound power to radiate light, X-rays, and high energy particles.[11] Viewed from a safe distance, the core will grow brighter than the galaxy that surrounds it such that, if you were so far away the galaxy had faded from view, the core would still be visible as a point of light. When a supermassive black hole dines on infalling matter, the resulting luminous object can outshine the entire surrounding galaxy. A quasar!

But it has sown the seeds of its suppression. This prodigious release of energy drives out gas and dust from the central regions, clearing them out of the raw material for black hole power. After 10 million years of shining brightly, the quasar will be starved into silence and its light will ebb to a modest level. The black hole will still be large and hungry, but just biding its time (Figure 9.6). After another hundred million years, it will have drawn in enough gas to begin another bout of activity. And the cycle will repeat.

The future of our galaxy as a quasar is speculation. Interactions and mergers do correlate with nuclear activity, but it's not deterministic: some mergers don't lead to quasars and some quasars don't seem to have been triggered by mergers.[12] If it does happen in our neck of the woods, the quasar phase will begin when the Sun's near the end of its life. Anyone around on Earth at that time will see the bright center of Milkomeda in the daytime sky, outshining any star or planet. Even in late middle age, the Galaxy has fire in its belly.

In our galaxy and elsewhere around the universe, the quasar phase is an event for the long-lived connoisseur. Creatures with short atten-

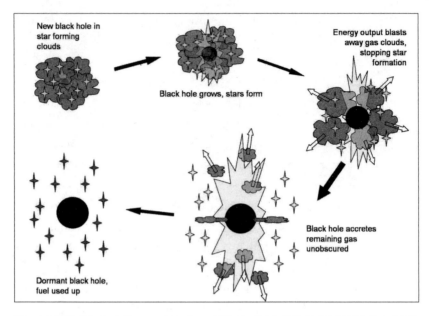

Figure 9.6. In this evolutionary scenario, a massive black hole at the center of a galaxy grows amid a bout of star formation, shrouded from outside view, until energetic outflows drive away gas and quench the activity. The black hole is starved of fuel and becomes quiescent, after which gas gradually falls to the center of the galaxy and another cycle of activity begins.

tion spans or short lives settle for entertainment from the violent death of stars or the gradually changing star patterns in the night sky. Only a persistent species that has mastered time can watch the ebb and flow of quasars across the cosmos and enjoy the grandiloquent statement their own galaxy makes when it activates its central beast: a light show for the ages.

Chapter 10

AGING OF THE MILKY WAY

The Milky Way has billions of stories, none too small to be told. Consider this: the tale of a carbon atom. Within all the billions of stars and their attendant planets in the Galaxy, carbon atoms are one in a thousand.

When the Galaxy forms from swirling clouds of diffuse gas, chilled by the expansion of space since the big bang, our atom is not on the scene. In a universe of primordial material, carbon can scarcely be imagined. Yet our atom forms in the Milky Way's youth, as a massive star cascades up the chain of nuclear fusion and turns three inert helium nuclei into a single carbon nucleus, ripe with promise.

The star is massive and has gorged on matter to excess; the churning of its interior sends our atom into its cooler outer layers, and then, late in its life, the star sloughs off sheaths and skeins of gas into the void of interstellar space. Billions of years pass, and the atom feels a slow but inexorable tug toward a nearby region of space where a young proto-stellar nebula is gathering itself. A young star forms, yellow and swollen; our atom watches from a safe distance. It is mere soot. In a delicate gravitational dance, it is incorporated into a rocky body 100 million miles from the star.

After billions of years of churn within the active core of the planet, something very strange happens. For a very short time—no more than the blink of an eye in the eons of the cosmos—our atom becomes a part of the complex transactions of the biosphere. Our atom is us.

The moment passes. The carbon atom is once more interred in rock and moved on sluggish conveyor belts of magma through the strata of the planet. Then liberation, belched into air by a volcano and, after a much longer span of time, further liberation as our atom, bound to two oxygen atoms, reaches the top of the atmosphere and bleeds into deep space.

Time loses meaning as the carbon drifts between stars, but eventually it feels a familiar tug and it enters for a second time the realm of stars. This star is smaller than the one that gave it birth and the atom is aloof from the mesh of interactions that consume the nearby particles. Eventually the star has nothing more to give. It settles into a compact form and our intrepid carbon atom is trapped in the collapsed core.

At rest, but not resting, our atom is shimmering with motion in the crystalline lattice of a white dwarf. Gravity will not relent; this is the final disposition of all who find themselves at the heart of a dead star. The atom will eke its last vestiges of heat into space until it has nothing to give but darkness: From soot to life to eternal black diamond.

Fade to Black

The Ebbing of the Light

Getting old sucks. We put a good face on it, but nobody can be truly equable as the body creaks and sags and the brain's edge is blunted. What about our galaxy? Does the Milky Way go quietly into the night, or, as Dylan Thomas exhorted his infirmed father, does it "rage, rage against the dying of the light." It's a little of the latter but mostly the former. The Milky Way in its Shakespearean sixth age finds "a world too wide for his shrunk shank" and is mortified as "his big manly voice turns again toward the childish treble."

We need to bring a new sense of time into view to talk about the fate of the Galaxy. It's about 12 billion years old now and we've followed it 4 billion years into the future, through merger with Andromeda, to the episodic quasar phase. What happens now plays out over 10 trillion years. If this span were a year, the age of the universe to this point would be just 10 hours. We're entering the realm of deep time.

Stars have been conjuring up the essential life elements—carbon, nitrogen, and oxygen—and unselfishly flinging them into space to become part of successive generations of stars and planets. It's an attractive idea: star death begets new life. As time goes by, our galaxy and all of the others get livelier and, let's dare to hope, smarter.

Unfortunately, nature disagrees. As the Milky Way ages, the cycle of star birth and star death will be irrevocably broken. Stars are made from the abundant gas that the universe contained when it was young. Now, nearly 14 billion years after the big bang, most of that gas has already been mopped up. Aging stars eject some fraction of their gas, and recycling enriches that gas with heavy elements, but as the eons pass, the flow of gas being recycled slows to a trickle. At some point, it will stop entirely.

Stellar evolution is counterintuitive. It seems that a big star should last longer than a tiny one. In fact, the converse is true. Large stars are spendthrifts, burning through their fuel in the blink of a cosmic eye, while small stars are misers, able to last a very long time on a small supply of hydrogen.[1] A star 10 times the mass of the Sun lasts 20 million years, less time than there have been whales in the oceans. A star like the Sun lasts 10 billion years. But a star just under a tenth the Sun's mass—the smallest a ball of gas can be and still be a star— can convert hydrogen into helium for 10 trillion years. It's a pathetic star, 10,000 times dimmer than the Sun. Even when fusion is done there's still heat in all that gas, however, so it takes 100 trillion years for a cooling red dwarf to fade into an invisible black dwarf.

Elliptical galaxies formed stars very efficiently early on so their gas is long gone; they steadily get older and redder. In spirals like the Milky Way, the terminal decline of star formation means there are no more stars massive enough to die as supernovae, leaving behind only dark neutron stars and black holes. Stellar cataclysms no longer seed space with the material to make new stars. Red dwarfs are the last stars left standing. For trillions of years they do yeoman work, eking out light from feeble nuclear reactions. (For a star, feeble is relative; the surface of a red dwarf is hundreds of degrees and its interior is millions of degrees.) And in 10 trillion years even the red dwarfs' guttering fires are extinguished all across the Galaxy.

The central, massive black hole is partially exempt from the stellar lockdown. With the gas all used up and no more being ejected, it can't dine on the food that makes it shine brightly. The quasar phase is over. But such a prodigious source of gravity can still pull in stars whose orbits carry them too close. (Our descendants are in no danger; the black hole at the Galactic Center is too far away to have any effect on the Solar System). The black hole continues to snack on stars and each time one enters the event horizon a distant observer sees a flash of light.

This is the end of the age of stars. It's as if the Milky Way is on a big dimmer switch, and nature is cranking down the power very slowly. Because our galaxy isn't special, the same thing will happen within each of the 50 billion other galaxies in the universe. The only exceptions to the rule of diminishing light will be galaxies in rich clusters, where mergers and the churning of gravity can keep star formation going a little while longer. Eventually, even clusters will turn into vast supergalaxies of dead and dying stars.

Various Corpses

We enter the age of stellar corpses. The universe has lost its shine. An inventory of the anticipated remnants shows roughly equal numbers of brown dwarfs and white dwarfs and a tiny fraction, only a few tenths of a percent, of neutron stars and black holes.[2]

The remnants of massive stars are dark. Nothing escapes the event horizon of a black hole and a neutron star is like an atomic nucleus with 10^{57} particles, its neutrons packed tight like eggs in a crate. A small fraction of neutron stars have hot spots on their surfaces that generate beams of radio emission. The stars spin, and in cases where the beam sweeps across the path of Earth we see a pulsar. It's not known exactly how nonuniformities in a neutron star crust generate radio emission so we can't predict how long pulsars will survive.

Half of the corpses were once massive enough to fuse hydrogen into helium like the Sun. They suffer a series of spasms and contractions and end their lives as white dwarfs. White dwarfs are stellar cinders, but their carbon-rich material is much denser than ash; it's crystalline like a bizarre form of diamond. The embers start off very hot but they cool rapidly. It takes only 100 million years for a 100,000°C (180,000°F) white dwarf to cool to 20,000°C (36,000°F), but then 800 million years to cool to 10,000°C (18,000°F) and 5 billion years to reach the Sun's surface temperature of 5500°C (9900°F). After that, the white dwarf continues to dim and slides through the spectrum from yellow to orange to pink and eventually a dull maroon. After trillions of years it is invisible, glowing only at invisible infrared wavelengths (Figure 10.1).

The other corpses never have a hot phase. Gas clouds that collapse

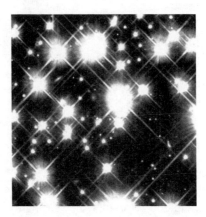

Figure 10.1. White dwarfs, in an image from the Wide Field Planetary Camera on the *Hubble Space Telescope*. The faintest stars in a cluster are 10 times less massive and hundreds of times fainter than the Sun. With no energy source, they eke out an existence as fading embers, Pink Floyd's "crazy diamonds."

to form objects less than a tenth the mass of the Sun can't light up the night. They never experience the glory of transmutation of elements. These failed stars give off dull and indistinct reddish light, hence the name brown dwarfs. Brown dwarfs are larger cousins of giant planets, and planets like Jupiter and Saturn also glow dimly in infrared waves. As eons roll by, brown dwarfs will eke their tepid heat into the utter cold of space and slowly fade to black (Figure 10.2).

Even as stars become old and diminished, gravity is still the engine of their activity. When they were young, gravity was an eager alchemist, forging elements, populating the periodic table, and making the sky blaze with Byzantine waste. When they get old, gravity turns into a vise, gently squeezing the gas and eking energy out into space. The Milky Way is still a majestic stage, but exits and entrances are long gone and the remaining actors are feeble and inactive.

Dissolution and Death

To go beyond the Dark Ages, our tour guide is Fred Adams, a physics professor at the University of Michigan with an avuncular, long-haired, and soulful presence. He points out that the Milky Way never becomes totally dull because gravity is still on the job. Adams has written many technical articles on star formation and cosmology and a popular book called *The Five Ages of the Universe* on the physics of eternity with his colleague Greg Laughlin.

He starts with a caveat. As the timescales move from the prodigious to the unimaginable, the predictions depend increasingly on poorly tested physics. Moreover, they depend on the assumption that physical laws don't change with time. A universe in which gravity and electric charge and the speed of light were all varying but their dimensionless ratio was the same would look identical to one in which they were constant. So experimentalists must look for changes in the most fundamental dimensionless number in nature, the fine-structure constant. It has a present-day value close to 1/137. There have been claims that the fine-structure constant changes with time, but the measurements are very difficult and the subject has become a graveyard for the reputations of experimentalists. We'll assume that nature is constant.[3]

Happily, even in the era of stellar corpses, the Milky Way is not totally dark. Left to its own devices, a dead star is dead. But we've

Figure 10.2. Brown dwarfs are "failed" stars, and as in this artist's impression, are thought to be surrounded by a disk of gas and dust that dissipates over time. Brown dwarfs emit heat, but temperatures are never high enough for nuclear fusion to occur.

neglected the fact that half of all stars are in binary systems. If they're in tight orbits, mass can transfer from one star to the other and cause a brown dwarf to put on enough weight to become a hydrogen-fusing star. The math is: dead + dead = alive. So mass transfer and occasional collisions between brown dwarfs provide a way for the Galaxy to keep forming stars 100 trillion years from now and beyond. The future Milky Way is a much diminished galaxy. Its army of 400 billion blazing stars will be replaced by less than a hundred dwarfs burning just above the hydrogen limit.

Stellar encounters also make occasional fireworks, which are note-worthy because they're superimposed on such a cold and endless night. A pair of white dwarfs can collide or merge to make a supernova explo-sion if the combined mass exceeds the threshold for violent detonation. This time, the math is: ember + ember = firework. An even rarer event sees a pair of neutron stars or a pair of black holes (or a neutron star and a black hole) colliding to emit an intense burst of high-energy radia-tion, the flash briefly outshining the rest of the universe. When these dense objects merge they also distort space-time and unleash a spasm of gravity waves (Figure 10.3).

The last fizz of stellar fusion is a sideshow in a circus completely run by gravity. In the era of stars, life was kept interesting by a battle of competing forces—radiation released from the creation of elements versus gravity. By 100 trillion years after the big bang, gravity may have lost some battles but it has won the war and has nobody left to play with but itself. Gravity playing solitaire turns out to be pretty interesting.[4]

Nature is parsimonious and likes to conserve energy. This leads to two effects in most gravitational systems. On the one hand matter tends to concentrate stuff toward the center. Gravity is always itching to cause collapse. But to conserve angular momentum some stuff gets flung out far from the center and some leaves the system entirely. This explains the Solar System, where a diffuse gas cloud collapsed and most of the mass went into a central object—the Sun—while a small fraction of the mass went into a set of objects at the periphery—the planets.

Figure 10.3. A supercomputer simulation of the gravitational wave radiation as it reaches its peak intensity during the merger of two black holes. The merged black hole is a small dot at the center. The LIGO (Laser Interferometer Gravitational-Wave Observatory) instrument hopes to detect gravity waves by the tiny distortion they cause in a pair of 5-km long metal rods suspended in a vacuum.

The same processes work on the scale of galaxies. We've already seen how the Milky Way and M31 merge to make an elliptical galaxy. That new configuration of stars is both more centrally condensed—stellar densities in the core of an elliptical are higher than in the present-day Galactic Center—and more dispersed—ellipticals are bigger than spirals. Now imagine two stars near the edge of a galaxy, both loosely bound so they almost have enough energy to leave the galaxy. If they pass near each other they'll exchange gravitational energy. On average one star loses energy and the other gains it. The star losing energy moves toward the center while the star gaining it is ejected. Galaxies can actually evaporate.

Evaporation and Ripples

Get a good book and take a seat because it'll be a long wait to see the Milky Way evaporate. This isn't going to happen for 10^{19} or 10 billion billion years. To get a sense of this number, let's rescale the analogy we used for the span of stellar fusion. If 100 trillion years were scaled back to one year, we'd be in the first 10 hours of that year. Now if we set the much larger time to the onset of the evaporation and collapse era to be the end of the day on December 31, and ask where we would be in this scaling 13.7 billion years after the big bang, it's one-thirtieth of a second after midnight on January 1, the very beginning of the year.

Gravity has one more trick up its sleeve. Gravitational radiation is to gravity what light is to electromagnetism. When a charged particle is accelerated it emits radiation, often light. When any massive body is accelerated it emits gravitational radiation, often called gravity waves. Gravity is so feeble that this radiation is imperceptible in the everyday lives of stars and galaxies. But it becomes important for the intense gravity of collapsed objects, or if we wait a very long time.

Since the emission of gravity waves causes a binary system to lose energy, the process is responsible for stars in a stable orbit gradually approaching each other and merging. Gravity waves are the reason that binary brown dwarfs will keep supplying the galaxy with a thin gruel of star formation long after fusion in single stars is exhausted. Gravity waves haven't been directly detected yet, but when Russell Hulse and Joe Taylor of Princeton University measured the decaying orbit of a binary pulsar, the results were exactly in accord with the prediction of gravity waves from Einstein's theory of general relativity. Hulse and Taylor won the 1993 Nobel Prize in Physics for their work. The same effect will also cause stars in the Milky Way to slowly spiral into the massive black hole at the center.

Keeping all these complex mechanisms straight would nonplus most physicists, but Fred Adams is unperturbed. He has a counterculture vibe, like a surfer or a savvy drug user, and he doesn't seem to take the serious business of the universe too seriously. But he does know how to weigh up all of gravity's tricks and here's what he concludes about the long-term fate of the Milky Way.

The emission of gravity waves will cause about 10 percent of the stellar corpses to spiral into the center of the Galaxy (similarly, leakage of gravity waves will be causing binaries to coalesce and planets to fall into their dead parent stars). The central black hole, currently "only" 4 million solar masses strong, will grow into a beast of 10 billion solar masses. Ninety percent of the dead stars will avoid the clutches of this monster because they'll be tossed out of the Galaxy. The dark matter will try to cling onto those stars but it can only slow and not stop the process of evaporation.

The future universe contains nothing resembling a galaxy, which we think of as a system of stars. Star husks drift in the great expanse of space, occasionally being vacuumed by marauding giant black holes. The night is exquisitely dark and only gravity speaks.

For anyone who has the patience to keep watching, the twin processes of black hole growth and stellar evaporation will play out on the largest cosmic scales. Just as the Milky Way and M31 merged and gobbled up the few dozen smaller galaxies of the Local Group, so the central black holes of groups and clusters and eventually even galaxy superclusters merge to produce a trillion solar-mass black holes that are the masters of millions of light-years each. Hierarchical clustering causes a cascade of structure on ever larger scales. In a reversal of Jonathan Swift's flea hierarchy, black holes always have bigger black holes to bite them.

And so the Milky Way fades from the scene, to a fate better than mere oblivion, but shorn of its bright baubles, "sans teeth, sans eyes, sans taste, sans everything."

Childhood's End

Futurology

"Never make predictions, especially about the future." So said Danish cartoonist Storm P, in a quote often attributed to Yogi Berra or Niels Bohr. We've seen that space travel is in its infancy and it would stretch our global resources to do something as modest as establish a Mars base. That would be our first baby step as a space-faring race. What might the future hold for a civilization that "comes of age" in the Milky Way?

We might one day learn that orcas and dolphins are just as smart as we are, with rich emotional inner lives and a transmitted culture and the awareness of their own mortality. But the corollary of their exquisite adaptation to the marine environment is the fact that their evolution has stabilized. Apes, meanwhile, turned their manipulative skills and precarious survival into an ascending set of cultural and technological masteries—fire, stone, iron, bronze, crops, machines, and now computers. It's notoriously difficult to predict the path of any recent innovation.

Consider these quotes from insiders relating to information technology. In 1943, Thomas Watson, the chairman of IBM, said, "I think there is a world market for maybe five computers." Six years later the hobbyist magazine *Popular Mechanics* offered this bold speculation: "Computers in the future may weigh no more than 1.5 tons." Here's an engineer at the Advanced Computing Systems Division of IBM in 1968 commenting on the microchip, "But what . . . is it good for?" Finally, listen to the icon of the information age, Bill Gates, from a lecture in 1981, "Who in their right mind would ever need more than 640k of RAM!?"[5]

Let's see how fast the crystal ball gets cloudy by the simple stratagem of looking at our own history in logarithmic intervals of time, and then reversing that to project into the future. In round numbers, there was no Internet a decade ago, a concept unimaginable to anyone under the age of 25. A hundred years ago there was no mass transit—no cars, no buses, and no airplanes. Travel by foot or horse was difficult and

FUTUROLOGY

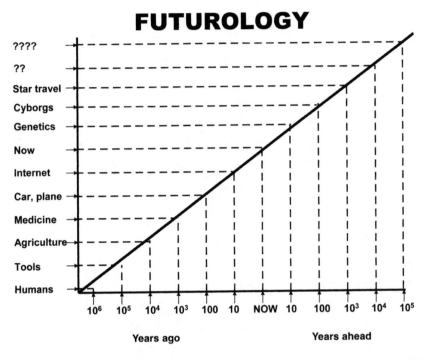

Figure 10.4. Looking at the past in logarithmic intervals, so that each interval is 10 times larger than the one before, quickly reaches us in a primitive state that would be horrifying and incomprehensible for someone from that time transported to our age. Similarly, the exponential progression of technology renders prediction very unreliable at any time more than a few hundred years into the future.

most people didn't travel far from where they were born. A thousand years ago there was no modern medicine and most lives were brutally short, truncated by diseases like tuberculosis and cholera that have little traction today. Ten thousand years ago, there were no cities and we lived as small bands of roving nomads. A hundred thousand years ago the human race was just getting started (Figure 10.4).

Now wind the clock forward. Ten years from now we imagine genetic engineering might have matured to the point of gene therapy and the control of diseases that used to decimate populations. A hundred years from now, we might project quantum computers that enable people to instantly access any knowledge via an ambient implanted Internet and hybrid human-machine entities or cyborgs. A thousand years from now it's not unreasonable to imagine travel to the stars.

But beyond that? It's too difficult to extrapolate. If humans survive 10,000 or 100,000 years, it's almost impossible to guess what capabilities we'll have. A million years is the mean longevity of a mammal line; will we have exempted ourselves from natural selection by then? In principle, there were Earth clones so early in the universe that human analogs could exist with a 10-billion-year head start. Will they have phenomenal capabilities, or will they peak and flame out or self-destruct much sooner?

Arks and Colonization

A long-term vision of our destiny in space involves travel to the stars. The Solar System is just our testing ground; once space tourism into low Earth orbit is routine and we've established colonies on the Moon and Mars, our eyes will naturally drift to a more distant horizon.

With that much real estate available, Earth clones and maybe even planets that are Edens beckon. Rocket pioneer Robert Goddard wrote a full technical proposal for an interstellar ark a century ago, but left it sealed in a drawer to avoid professional ridicule. Space colonies and interstellar arks quickly became science fiction staples; many people are familiar with the images of living worlds in the shape of cylinders or wheels where rotation provides artificial gravity. The prospects for constructing them have improved with new propulsion technologies and lightweight materials, leading to some unabashed optimism.

Here's what former NASA Administrator Michael Griffin said in 2005 about humans living beyond Earth: "The goal isn't just scientific exploration, it's also about extending the range of human habitat out from Earth into the Solar System as we go forward in time. . . . I don't know when that day is, but there will be more human beings who live off the Earth than on it. We may well have people living on the Moon. We may have people living on the moons of Jupiter and other planets. We may have people making habitats on asteroids. I do know humans will colonize the Solar System and one day go beyond."

Not everyone is so upbeat. Here's what science fiction writer

Charles Stross had to say on the same topic: "In the absence of technology indistinguishable from magic—magic tech that does things that from today's perspective appear to play fast and loose with the laws of physics—interstellar travel for human beings is near-as-dammit a nonstarter. And while I won't rule out the possibility of such seemingly-magical technology appearing in the future, the conclusion I draw *as a science fiction writer* is that if interstellar colonization ever happens, it will *not* follow the pattern of historical colonization drives followed by mass emigration and trade between the colonies and the old home soil."[6] He was taken to task by readers because they consider it the duty of a science fiction writer to be an optimist about space travel or colonization. Who's right—the gung-ho bureaucrat or the gloomy futurist?

One of the biggest hazards for interstellar colonizers isn't collisions with space dust or meteors, but hits from subatomic particles called cosmic rays. The highest-energy cosmic rays are protons traveling a shade slower than light; they have as much kinetic energy as a well-hit tennis ball yet the particle involved is 10^{15} times smaller. Only a cosmic accelerator can pump a proton to a million times the energy reached by the Large Hadron Collider. Astronomers speculate that these protons get accelerated by supermassive black holes in active galaxies like quasars. The energy tied up in cosmic rays rivals all the energy in starlight.

The effect of very high energy cosmic rays on human tissue would be profound—think of a Major League fastball delivering all of its energy not to your bat or even your elbow but to a tiny region within a vital organ. On Earth, we're multiply protected. Many are blocked as they encounter the heliosphere, a bubble of space created when the solar wind and magnetic field hit the diffuse gas of interstellar space. The magnetic field of Earth intercepts and redirects many more. Last, the rest are stopped in their tracks by the atmosphere. At these high energies, one cosmic ray passes through any square meter of space each year, which amounts to a lot of impacts on a long trip between stars. The forward velocity of the colonizers' spacecraft increases the damage.

Another technical issue to overcome is suspended animation. The nearest stars are light-years away and because travel at anything more than a few percent of the speed of light is implausible, a one-way trip to anywhere interesting would take decades or centuries. The people on cryonics use an antifreeze solution and chilling by liquid nitrogen but the survival rate is currently unknown. Daring doctors have used hypothermia and saline solution to put dogs and pigs into a metabolic state akin to suspended animation and revive them after a few hours with a 90 percent success rate. Another approach that's worked with mice substitutes hydrogen sulfide for oxygen and doesn't require freezing. But working with animals and achieving nominal success after a few hours is a baby step compared to reversible suspended animation on humans that works reliably over many years.

Interstellar travel will be unavoidably difficult due to the vast distances involved. The nearest stars are hundreds of thousands of times farther away than Mars and the nearest Earth-like planets will be millions of times farther away. Assuming a minimum life support package of a one-metric ton space capsule, holding one voyager and hardened to resist cosmic rays, it would take 2×10^{19} Joules of energy to get it to Alpha Centauri at 10 percent of the speed of light using an engine that can convert energy into momentum with 10 percent efficiency. It would usurp the world's entire energy usage for two months just to send that one traveler.

Difficult, and expensive. The *International Space Station* is 10 years old, three-fourths complete, and its eventual cost will be $100 billion. It supports a crew of six, but it needs to be constantly resupplied and it functions like a cramped hotel room with no room service. It's only in low Earth orbit, equivalent to an afternoon drive to the countryside, straight up.

So forget the "space wheels" with smiling colonists living and playing as they approach their new home. Looking at some of NASA's designs from the 1970s makes you wonder what was in their Kool-Aid (Figure 10.5). The first interstellar travelers will be sealed into metal "coffins," their metabolisms lowered to within an iota of death, hurled into an

Figure 10.5. In the 1970s, still giddy from the success of the Apollo missions, NASA commissioned artists to visualize space colonies with artificial gravity that could house 10,000 people in total comfort. The artwork is a poignant reminder of how hard it really is to travel, work, and play in the unforgiving environment of space.

utterly cold void to cross the Galaxy and face an unknown future. Only the brave and the insane need apply.

Send in the Bots

If protecting the physical and psychological frailty of people is a tall order, machines may hold the key to our endless dominion over the Galaxy. With machines we can ride the wave of miniaturization that packs more sophistication in a smaller package each year. It's not a complete stretch to imagine a propulsion system, powerful computer, remote sensing camera, and communication system in a package the size of a baseball. The energy requirement would be 10,000 times less than the space coffin just described.

With economies of scale in production we might be able to afford to send out hundreds of bots, fanning out across the Galaxy at a tenth the speed of light, reporting back on distant solar systems to us and our descendants. We could overcome the long travel times by numbers, such that on average one probe would give us a view of new worlds every month. It would be the ultimate reality show.

The idea becomes compelling if the space probes were able to find asteroids or small moons, mine materials, and construct replicas of themselves. Then if we send out a dozen probes, that dozen make a dozen more each, and the fleet of explorers grows exponentially. At 10 percent the speed of light, it would only take a few million years to explore the entire Galaxy. The problem has been "reduced" to constructing a self-replicating space probe.

That's by no means trivial. We're just beginning to develop printers that can make endless three-dimensional copies of solid objects from plastic or metal. In 2008, Adrian Bowyer, founder of the RepRap project, programmed a machine that created an entire set of parts for an identical daughter machine. But then it just sat there—it couldn't assemble the daughter. The futurist Eric Drexler calls such machines "clanking replicators." He's written about the other end of the size spectrum, nanotechnology that he thinks will soon be used to make micromachines one atom at a time. It's a long road from RepRap to a machine that can mine and refine raw materials, use those materials to make components, and then assemble those components into working sensors, computers, propulsion systems, and cameras.

The theory of universal replicators was developed in the 1950s by the computer science pioneer John von Neumann, and as a result self-replicating space probes are often called von Neumann machines. But von Neumann was interested in something more profound—the set of instructions that described a general computational device that could be used to compute anything.[7]

How long before we build self-replicating space probes? Hundreds, maybe thousands of years. We've taken the first baby steps already. In the development of intelligent life, that's a small increment, so if we're on the verge of doing it we can imagine that someone else got there first. Physicist Paul Davies thinks it is quite plausible that this has already happened: "The tiny probes I'm talking about will be so inconspicuous that it's no surprise that we haven't come across one. It's not the sort of thing that you're going to trip over in your backyard. So if that's the way the technology develops—namely, smaller, faster, cheaper, and if other civilizations have gone this route—then we could be surrounded by surveillance devices."[8]

The idea of tiny probes that passively observe us and beam back information to the home planet is creepy but benign. However, self-replicating probes could easily be programmed to grow in size and capabilities until they become dominant and destructive. A programming slip or evil intent is all it would take. Maybe the mistake will be ours and a robot designed to terraform Mars will malfunction, reproduce, and run amok. When the machines move beyond replication to evolution they take on attributes of life, a speculation that's familiar from the *Terminator* movies. Fred Saberhagen concocted a space opera around the idea of "berserkers," robotic aliens intent on seeking and destroying all forms of organic life. Maybe it's better if we're alone in the universe.

The Great Silence

In 1950, Enrico Fermi was having lunch with two colleagues at the University of Chicago. Fermi would later win a Nobel Prize in Physics for his work on fission. He was known as "the Pope" as a nod to his infallibility on anything scientific. They were joking about newspaper reports of local UFO sightings when he mused, "Where are they?"

Fermi knew that the size and age of the universe and the mediocrity principle favored the existence of lots of extraterrestrial life. He also knew that human technology was young, so there were likely to be a lot of alien civilizations who could easily communicate or travel across large distances in the Galaxy. And yet, UFO reports aside, there's no sign that anyone has visited and no evidence of messages or robotic emissaries from alien civilizations.[9] Since 1950, the relevance of his question has only grown. The Search for Extraterrestrial Intelligence (SETI) has been operating for 50 years with constantly increasing sensitivity and bandwidth but it has been met with what researchers call "the Great Silence."

How should we interpret the Great Silence? There are many possible answers, as there are to Fermi's question. There are likely to be 100 million habitable terrestrial worlds in the Milky Way, and they are an average of 1.5 billion years older than Earth. There's plenty of time and real estate available for developing complex, intelligent life. The earliest biology experiments in our galaxy might have started 10 to 11 billion years ago. For us to see no trace of intelligent life is a puzzle that begs for an answer.

Our exponential rate of technological progress means we should take Fermi's question seriously. Even if self-replicating probes are to our technology as our technology is to hunter-gatherers, that's still the near future, perhaps tens of thousands of years away. Technological progress is extremely rapid compared to biological progress. Probes like these could seed life on suitable planets in a mechanical form of panspermia; they could act as sentries to monitor the emergence of intelligent life; and they could explore more efficiently and safely than intelligent organisms could.

A fleet of robotic probes could also act as the nodes of a network of interstellar communication—think of them as "Encyclopedia Galactica." Follow the exponential increase in computer power and speed, called Moore's law, another few decades and we'll be able to put 10 terabytes on a flash card. That's enough for a video log of a human life: everything you see, say, hear, and experience recorded digitally. Project exponential progress in computing another 50 years and we will have the ability to record the video logs of a century of our entire civilization—10^{23} bits—onto a kilogram of atom-scale storage, which will probably be based on addressing carbon atoms in a lattice. Think of it as our message in a bottle. Or anyone else's.

Explanations for the Great Silence are often contorted. They exist but make a cultural choice not to explore or communicate. They exist but are observing us and not making themselves known: the "zoo" hypothesis. They're waiting for us to mature to either greet or destroy us: the "berserker" hypothesis. They are so advanced we wouldn't know how to recognize them: the neophyte hypothesis.

There is of course a simpler explanation for the Great Silence: they don't exist. We might be unique or extremely rare as a technological civilization. One or more of the probabilities that plug into the Drake equation (discussed at the end of Chapter 4) could be much lower than the generally optimistic values assumed by astronomers. The formation of life from simple chemical ingredients might be a fluke. The transition from microbial life to the level of complexity of large animals may be a highly contingent result of evolution. Or technological "maturity" may be a bottleneck due to the instability that goes with having enormous power over nature.

It's consistent with everything we know that we're the first spacefaring race in the Galaxy. Perhaps the first, last, and only. Knowing that we're alone would be both exciting and intimidating. Hopefully it would also encourage us to navigate our childhood successfully so we could enjoy the space and bounty of the universe beyond our home.

Chapter 11

HOW THE UNIVERSE ENDS

In the beginning there is nothing. That is to say, there is no thing, no tangible substance, but there is the potential to make anything, all things. Nothing is too disparaging a word for such opulent possibility.

It's not a place, because there is no up or down, no in or out. It's not a time, because there are no beginnings or endings. Vacuum conveys the absence well enough, but the word doesn't capture the richness of the energy states in a situation when quantum indeterminacy means that rules can be broken, and if they're broken fleetingly, the transgression can be extreme.

In the seething quantum foam, bubbles of space-time come into being and disappear. Most are evanescent, but a few borrow enough energy from the vacuum to inflate and endure. These bubbles are diverse. Some last a fraction of a second; some last eons. Some are smooth and featureless; others contain a chaos of black holes. Some are governed by light; others are governed by matter. Some contain time; others are timeless. Some are no bigger than an atom; others are unbounded. This is the multiverse.

In one of these bubble universes, conditions are "just so," so we enter a just-so story where the storyteller has a knowing smile and a twinkling eye. In the special universe where the bed is not too hard or soft and the porridge is not too hot or cold, a small skewness in the laws of nature causes there to be enough substance to make 10,000 billion billion stars and enough time for those stars to forge heavy elements and shelter complex but fragile forms that emerge when those elements combine. Is a universe that contains self-awareness aware?

The fragile forms are part of the universe yet apart from it. They are obsessed with value and meaning, with purpose and intention. They feel as contingent and evanescent as the bubbles that gave rise to all they know.

It is not about them. It was never about them. The universe has gone from unimaginable, featureless heat to complexity and it will return in time to unimaginable, featureless cold. This is their time. This is our time.

Something from Nothing

The Expanding Universe

Finally we turn to the grandest stage of all. The universe is defined as everything that exists: space, time, matter, energy, and the physical laws that govern their behavior. Deriving a detailed, predictive theory of the universe is perhaps the greatest intellectual challenge humans have ever undertaken.

A hundred years ago, the conventional view held that our galaxy *was* the universe. Estimates of distance were unreliable due to a sprinkling of obscuring dust in between stars, so estimates ranged from 10,000 to 100,000 light-years. In his mapping project, William Herschel had cataloged hundreds of fuzzy objects called nebulae and they became the subject of controversy. Most astronomers thought they were proximate regions of star formation in the Milky Way. But others had revived a radical idea of the German philosopher Immanuel Kant that the nebulae were distant star systems like the Milky Way, or "island universes," whose individual stars were too faint or close together to be seen by telescopes.

The issue was resolved by Edwin Hubble, a young man who oozed the self-confidence you might expect from somebody who was a Rhodes Scholar, a boxer, an athlete, and an attorney before turning his hand to astronomy. In the 1920s he used the newly constructed 2.5-meter telescope at Mount Wilson near Los Angeles to show that faint variable stars in Andromeda behaved identically to similar variable stars in the Milky Way. The relative brightness of the Andromeda and Milky Way variable stars told him Andromeda was a million light-years away, far beyond the edge of our galaxy. Soon he had observed several dozen galaxies and the most distant were 40 or 50 million light-years away.

He wasn't done yet.[1] Using spectra of these galaxies, he showed that features in the spectra were shifted to the red by an amount that was proportional to the distance. This linear relation between distance and recession velocity is called Hubble's law. Its simplest interpretation is that the universe is expanding. Each galaxy is a vast stellar system of many billions of stars and they're receding from us at millions of miles per hour and the furthest ones are moving away the fastest. Hubble's discoveries placed us in a vast and dynamic universe (Figure 11.1).

This new information raises many questions, and the observationally minded Hubble didn't even try to answer them. What is expanding? Why is it expanding? What is it expanding into? And if everything is moving away from us, doesn't that put us back at the center of the universe, subverting the great idea of Copernicus?

Albert Einstein had provided the theoretical apparatus for interpreting Hubble's data just a decade earlier. General relativity is his tow-

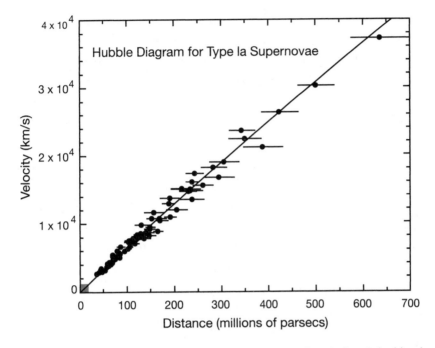

Figure 11.1. There is a linear relationship between cosmic expansion rate (here in km/s) and distance (here in millions of parsecs, where a parsec is about 3 light-years) for all galaxies, meaning that distant galaxies move apart faster. Linear expansion projects to a time in the past when the galaxies were all together. The range of Hubble's original measurements is shown by the small shaded box near the origin of the graph.

ering theory of gravity, and its conceptual basis is totally different from the linear space and time of Isaac Newton. In Einstein's theory, space and time are coupled and there's a mathematical relationship between the curvature of space-time and the amount of mass and energy. This is true on the small scales of compact objects like black holes and on the large scales of the universe. The math of general relativity is fiendishly difficult but to its practitioners the theory is both elegant and beautiful. However, it predicts phenomena that tax intuition and the imagination.

The easiest question to address is the concern of chauvinism. If all the galaxies appear to be moving away from us it seems like we must be at the center of the expansion. But no, it's a question of perspective. Imagine astronomers situated in a galaxy a million light-years distant, observing the Milky Way and other galaxies. They would see the Milky Way moving away from them at the same speed that we see them moving away from us, and they would see galaxies in all directions receding with speeds that increase with increasing distance, placing them on a Hubble law (maybe they have their own famous astronomer who discovers this so they give the law someone else's name). Observers on any galaxy see the same recession of other galaxies. They can't all be at the center of the universe; in fact, none are.[2]

Answers to the other questions are less satisfying because they speak to the strangeness of general relativity. Galaxies are not like shrapnel in the aftermath of an explosion. They move apart due to expanding space-time. Space-time is an entity describable only by mathematics. Because space-time expands it's not necessary for it to expand "into" anything and so the universe is self-contained and needs no external container. As for the cause, we learn a lot by tracing the expansion backward in time, to when the universe was a very different place.

A Day Without a Yesterday

If galaxies are getting farther apart every day, the linear expansion can be projected back to a hypothetical time when everything was on top of everything else. This time is the age of the universe. A linear projection of Hubble's law backward overestimates the age because gravity works to oppose the expansion and so the expansion rate was faster in the past.

Hubble's law says that the universe hasn't always been this large and empty. It was once much smaller and denser, and—like any gas when it's compressed—hotter. By the 1940s physicists had speculated that the initial state of the universe was one of infinite temperature and density, an iota of highly curved space-time and compressed mass-energy. Georges Lemaître, a Belgian theorist who was also a Jesuit priest, called it a "day without a yesterday," and the Russian émigré George Gamow calculated that there should be a dim afterglow from creation, present everywhere in space. All that was left was a catchy name for the theory. That was provided by English cosmologist Fred Hoyle, who preferred a rival theory. He called it the big bang and his attempt at disparagement backfired because the moniker is memorable.

Figure 11.2. The sky as observed by the NASA's *Wilkinson Microwave Anisotropy Probe.* The overall temperature of the radiation is 2.7 Kelvin (−454°F) and the tiny variations are quantum fluctuations dating from the first tiny fraction of a second after the big bang. This radiation affirms the origin of space in the hot big bang.

The big bang model rests on a sturdy tripod of evidence. First is the Hubble expansion, recently confirmed as being linear to a distance of over a billion light-years. Second is the cosmic microwave background radiation, that tepid relic of the big bang predicted by Gamow. It was detected accidentally at Bell Labs in 1965, coming with almost equal intensity from every direction in the sky, at just 2.73 degrees above absolute zero on the Kelvin temperature scale (−454°F). Third is the abundance of light elements in the universe. About a quarter of the universe by mass is helium, which is too much to have been made by stars, but the right amount to be explained by the big bang—that helium was forged when the infant universe was a few minutes old and as hot as the core of a star like the Sun.[3]

Recent observations have served to bolster the big bang and push the horizon of our knowledge ever closer to the singularity of birth. The smoothness and flatness of the universe have no ready explanation in the standard big bang so the idea was extended to include extremely rapid, exponential expansion when the universe was an unimaginable 10^{-35} seconds old. The inflationary big bang model has received some tentative support from subtle signatures in the microwave background radiation seen by the *Wilkinson Microwave Anisotropy Probe* (Figure 11.2).

As we trace the big bang back to the origin, we reach the frontier of knowledge of theories of matter. The physics of the big bang is well tested back to an age of a microsecond, when asymmetries in forces led to the preponderance of matter over antimatter. Before that the big bang exceeds the abilities of any accelerator on Earth. It points to a time when the four forces of nature were melded into a superforce. Current theories are inadequate to understand the origin because they can't unify the large-scale gravitational force with the small-scale forces that govern quantum behavior.

A snake that eats its tail is called an ouroboros; it's one of the oldest symbols in human culture. The ouroboros is a perfect metaphor for the unity of the very large and the very small represented by the big bang. Humans are poised at the logarithmic midpoint of this range of scales. We dare to contemplate both extremes of existence.

The big bang model is audacious. It's amazing that astronomers can hide their smirks when they describe it. It says this vast and mostly empty universe emerged from an infinitesimal point of space-time of infinite density and temperature.[4] It says that the seeds for galaxy formation resulted from quantum ripples blown up to a macroscopic scale by inflation. It says the energy for the expansion may have been borrowed as a quantum transaction from the potential of the vacuum. So 10^{80} particles and 10^{89} photons essentially came from nothing.

A Vast and Ancient Void

A century of breathless progress since Hubble has acquainted us with a cosmos 13.7 billion years old, its mostly empty space populated with about 50 billion galaxies. The cozy universe of antiquity—just a million miles across in the ancient Greek version—has been pried open to an unimaginable size.

The heliocentric model set the fixed stars a billion miles away. In the nineteenth century astronomers measured the distances to stars for the first time and the universe got 100,000 times bigger, hundreds of light-years, or 10^{14} kilometers. Herschel's mapping of the Milky Way added another factor of a thousand, to 100,000 light-years, or 10^{17} kilometers. By the time Hubble was done, he had expanded the universe, literally and metaphorically, to 100 million light-years, or 10^{20} kilometers. Since then, with successively larger telescopes, our reach is a hundred times farther still. Miles or kilometers, take your pick, 10^{22} is a very large number.

We can put this number within range of our intuition with a set of two scale models. Shrink the universe by a factor of 300 million. Earth reduces to the size of a golf ball, fitting comfortably in your hand. The Sun becomes a glowing three-meter ball 400 meters away and the Solar System is the size of a small town. But the nearest star in this model is 35,000 kilometers (20,000 miles) away so most of the universe is still beyond view. Now shrink by a second factor of 300 million. The stars reduce to microscopic scales and the typical distance between stellar systems is three millimeters. The Milky Way is a twisting spiral about 300 meters across and the nearest comparable galaxies are about a kilometer distant. Even with a factor of 10^{17} reduction in scale, the most distant galaxies are 45,000 kilometers (30,000 miles) away.

Understanding the universe's scale starts with a conceptual problem well known to the ancient Greeks: if the universe has a finite size and an edge, what lies beyond the edge? But if it has no edge how can we conceptualize something that's infinite? There's also the practical issue of measuring space, which is invisible. We see objects that exist in space, but distant objects are extremely faint and perhaps beyond the reach of our telescopes.

Measurements are complicated by cosmic expansion and by the finite speed of light. When we ask how big the universe is, we presumably mean how big the universe is *now*. But light doesn't travel instantly through the universe; we see distant regions as they *were*, not as they *are*. So when we observe distant regions, we're looking back to a past epoch when the universe was smaller. Size and age have to be derived from a model for cosmic expansion since the big bang. Fortunately the model's parameters are well determined.

Hubble discovered that more-distant galaxies are moving away faster. At some point, we're looking out so far that the galaxies we see were moving away from us at the speed of light at the time their light was emitted. That distance is the size of the observable universe. Photons from those galaxies act like Alice in the Red Queen's race in *Through the Looking Glass*, running as fast as she can to stay in one place. If the age of the universe is 13.7 billion years, you might think that the edge of the observable universe in any direction is 13.7 billion light-years away. But it's actually much larger, because for most of the universe's history the expansion rate has been decreasing as matter tugs against the initial propulsion from the big bang (Figure 11.3).

Looking out in space is looking back in time. A galaxy 5 billion light-years away is like Alice in the Red Queen's race. It was moving away from us at light speed long ago but since then its recession speed has

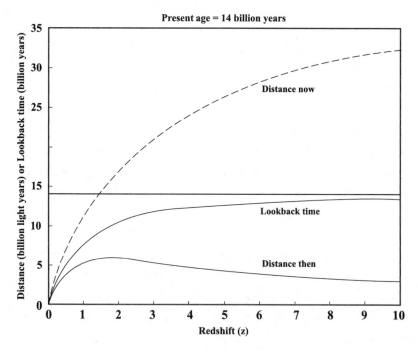

Figure 11.3. A graph showing some of the bizarre and counterintuitive consequences of the fact that distant parts of the universe can recede from each other faster than the speed of light. As redshift increases, lookback time approaches the age of the universe, 13.7 billion years. The lower curve shows the distance in billions of light years (Gly) or the lookback time in billions of years (Gyrs) between any two points at that redshift at the time the light was emitted. The upper curve shows the distance between any two points at that redshift now.

diminished, so its photons can catch up to us after traveling immense distances. Our largest telescopes can actually see even more distant galaxies. Their light is 12 billion years old, and when they emitted it, they were receding from us at twice the speed of light.[5] Thanks to this extreme stretching of space, the limit of our vision, and so the size of the observable universe, is about 46 billion light-years in all directions.

Does space stop there? No. Recall that the big bang model contains an added ingredient called inflation, which is needed to explain the overall smoothness and flatness of space. Inflation expanded the universe by a huge factor, such that our observable universe is just a tiny patch of a much larger region of space-time.[6] We don't know how much larger because the inflation model doesn't specify the underlying space-time. The patch that became our universe was a quantum fluctuation in the model and the nature of that fluctuation is inevitably uncertain. But if inflation occurred, the physical universe—all there is—is much larger than the observable universe—all we can see.

The Shape of Space

The universe's shape is much easier to measure and understand than its size. General relativity says the universe has a global curvature set by its average density of matter. Depending on how much mass there is, the universe might be flat, positively curved like a three-dimensional version of the surface of a balloon or negatively curved like a three-dimensional version of the surface of a saddle.

That's the concept, but what about the practical problem of measuring the shape of something you can't see? Astronomers tried to measure the shape of space for decades using galaxies and even quasars, without success. The answer came with careful observations of the microwave background radiation, that sea of low-energy photons leftover from the big bang. We see slight temperature variations, or "speckles," in those microwaves across gulfs of time and space because the radiation dates from a time when the universe was only 380,000 years old.

Think of the universe as a gigantic optics experiment. If space had a positive or a negative curvature, the speckles would be magnified or demagnified relative to the situation of flat space. In fact, we see the speckles with just the size they would have if space were flat, with no magnification or demagnification. The universe therefore has almost no curvature. Recent measurements with the *Wilkinson Microwave Anisotropy Probe* have extraordinary precision and set the departure from flatness to be less than 2 percent.

However, the fact that the visible universe is a subregion of a much larger space-time opens up interesting possibilities. In the language of general relativity, our local geometry is flat (if you can call 92 billion light-years local), while the global geometry is not easily measured. The global geometry might be a continuation of local geometry—flat space extending infinitely in all directions. But it might be much more interesting and exotic; a variety of positively and negatively curved geometries is possible. In fact, we must make a distinction between the curvature of the space and its topology. Topology includes other mathematical ideas like connectedness—whether you can actually get from point A to point B, or even whether if you can get from A to B you can get back from B to A!

The best handle we have on the global geometry is the microwave background radiation, but its imprint on that radiation is very subtle, so there has been controversy and no consensus about any of the claims for departures from flat space. In the past five to six years, different groups have claimed to see evidence for global geometry in the shape of a cylinder, a torus, a soccer ball, and a horn. Remember that these familiar objects are curved in two dimensions; for the situation of the universe, you have to imagine the curvature in three dimensions. Yes, it's hard. The practitioners park their imaginations at the door and must abandon themselves to fearsome math to understand curved space.

The first three are all examples of positively curved global geometry. Those spaces are finite so radiation can potentially travel through the universe multiple times, and the signature is patches of opposite sides of the sky where microwave speckles match, analogous to the patches on a soccer ball (a soccer ball is one of a small set of highly symmetric spaces, formed by seamlessly joining pentagons). The last example is a negatively curved global geometry, called a Picard horn (named after a French mathematician, not a reference to the fictional captain of the *Enterprise*). The signatures of this type of universe are an absence of very large patches of structure, and small patches that are elliptical in shape rather than circular since the negative curved space acts like a distorting lens.[7] A sensitive new measurement is expected from the European Space Agency's Planck satellite, launched in 2009.

The Future of Structure

It's hard to imagine the universe will end because it seems eternal and unchanging. The expansion of the universe is imperceptible on human timescales. Astronomers have pieced together the story of how cosmic structure formed and are able to make predictions for the future. The most interesting attribute of the universe is its lumpiness. Without that there would be no stars or galaxies, and without those there would be no planets or people!

The history and future of the expanding universe are governed by four major components: photons, normal matter, dark matter, and dark energy. There are 2 billion photons for every particle, and during the early fireball phase the universe was an almost perfectly smooth soup of high-energy photons and particles moving close to the speed of light. At 55,000 years after the big bang the densities of matter and radiation were equal. As the universe expanded, the density of matter dropped due to the growing volume, but the density of energy dropped faster because each photon was stretched or redshifted to lower energies by the expansion. Radiation hasn't been influential since the universe's childhood.

With the force of radiation ebbing, gravity got a grip and started to form cosmic structure. About 380,000 years after the big bang, the temperature was 2730°C (4950°F), cool enough for stable atoms to form. But it was 100 million years before the lights went on in the universe as the first stars formed. Because there was no carbon yet we can be pretty sure nobody was around to catch the pretty light show.

The great project of galactic construction was governed by something called dark matter, which outweighs normal matter by 6 to 1. Over the past three decades astronomers have accumulated a lot of indirect evidence for the existence of matter that doesn't emit radiation of any kind. It can't be black holes, dead stars, free-floating planets, or rocks or dust particles in space, which leaves only subatomic particles that interact weakly with normal matter. Physics theories provide guidance as to what the dark-matter particle might be, and several experiments are underway to directly detect this major component of the universe.

Gravity formed small concentrations of dark matter first, and then larger structures formed from the merger of smaller structures, in a bottom-up hierarchy. Normal matter fell into the center of these dark matter gravity "wells." The earliest galaxies formed about 300 million years after the big bang. Small galaxies formed first, then merged to form larger galaxies, and finally superclusters of galaxies that are only converging by gravity now for the first time. We see traces of merger processes in our moderately large galaxy which has consumed lots of dwarf galaxies to reach its current size. The traces are spaghetti-like streams of stars in the halo.

Counterintuitively, star formation within the largest galaxies peaked first, about 3 to 4 billion years after the big bang, and star formation in the tiniest galaxies 7 to 8 billion years after that. The age of galaxy and star formation is now mostly over. Those fires burned 10 times more brightly in the first three-quarters of the history of the universe than they do now. The expansion of the universe has helped to quench the formation of new structures; the gas that might be available to form new stars is marooned in intergalactic space.

For most of cosmic time, dark matter acted as a brake on expansion. But about 5 billion years ago, a new and surprising component of the universe took over: dark energy. A pie chart of the universe shows it's made up of about 74 percent dark energy, 21 percent dark matter, and 5 percent normal matter (Figure 11.4).[8] Amazingly, most of the normal matter is hot and diffuse and in the space between galaxies—the material that makes up 10^{22} stars in 50 billion galaxies is just 0.5 percent of the universe!

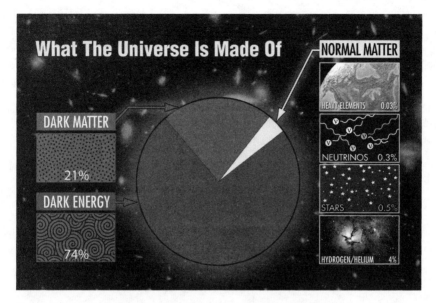

Figure 11.4. Pie chart of the contents of the universe. Normal matter is only 5 percent and the sum of 10^{22} stars is only 0.5 percent of the total. Neither dark energy nor dark matter is well-understood but they completely dominate the expansion of the universe.

How It All Ends

Dark Energy

Saul Perlmutter was nervous. The Lawrence Berkeley Lab physicist had spotted a signal in his supernova data that shouldn't have been there. In fact it had the opposite sign to the signal he and everyone else on his team expected. Cosmology at the observational limit is difficult so maybe the data were too noisy to be reliable. Maybe he'd fallen prey to one of the systematic errors that snake-bite scientists from time to time—even the great Edwin Hubble had overestimated the expansion rate of the universe and underestimated its age by a factor of 7 due to an unrecognized systematic error.[9] The private, dark thoughts may have unsettled Permutter the most; he was an outsider in the game of cosmology and some were waiting for him to stumble.

But another team led by Brian Schmidt and Adam Reiss at Harvard University had the same result. In 1998, both groups published their data and landscape of cosmology was irrevocably changed. They were pursuing Hubble's agenda of using objects of known brightness to trace the history of cosmic expansion. The variable stars used by Hubble can only be seen out to 3 billion light-years so beyond that a new tool is needed. Both teams favored the type of supernova that forms in a binary star system, because each explosion has a fixed peak brightness. The supernova rivals the brightness of the galaxy that contains it, so it can be seen to very large distances.

What did they see? The standard model of a decelerating expansion predicted how faint a supernova should appear at a distance of 5 to 6 billion light-years. The supernovae both teams observed were 30 percent fainter than they should have been, on average. Either the exploding stars were strange and intrinsically dim for some reason or they were 15 percent further away than they should have been. The explanation was that space had somehow expanded by that much extra relative to a universe where the expansion was controlled by matter. The extra "something" was causing the expansion to accelerate.

More than a decade later, we still don't know what that something is. Like "dark matter," the phrase "dark energy" is more a container for ignorance than a physical description. All we really know is that there is a force that behaves like antigravity or gas with negative pressure, a force that started to dominate the behavior of the universe about 5 billion years ago. The universe has a brake and an accelerator. The brake is the dark matter and it's been getting weaker as the universe expands and the cosmic density drops. The accelerator—dark energy—has always been present and its effect is now dominant. We're stuck in a runaway universe.

Saul Perlmutter would have been *really* nervous if he'd known how much angst dark energy would create in the physics community. In the early 1920s, before Hubble's great discoveries, Einstein was told by astronomers that the universe was static, yet his theory favored a dynamic universe. He "doctored" the solutions to his equations with a cosmological constant to counter their tendency to predict a collapsing universe.

The dark energy could be Einstein's cosmological constant, or it may be an exotic kind of energy that varies over space and time, called quintessence. Standard physics ascribes energy to the vacuum, but the best calculation predicts energy 10^{120} times larger than needed to explain cosmic acceleration. We could jettison general relativity, which is very successful as a gravity theory, or reach for bold new physics. It's hard to say who should be more embarrassed, the astronomers who can only account for 5 percent of the universe, or the physicists who miss the mark in estimating dark energy by 120 orders of magnitude!

Regardless of our theoretical confusion, the observation stands. The universe is expanding at an ever-increasing rate. If this continues it promises a very lonely ending. But a caveat: The properties of dark energy are almost unconstrained by either observation or theory so predictions of long-term outcomes should be taken with a pinch of normal nondark matter.

Lost Horizons

In the expanding universe not everything expands. Your body and car and house aren't expanding because they're held together by atomic forces. The Solar System isn't expanding because it's held together by the Sun. Galaxies, including the Milky Way, aren't expanding because the stellar orbits are bound by an encompassing well of dark matter.

Dark energy will pull many galaxies away from each other but it won't violate the structural integrity of clusters of galaxies or groups of galaxies, at least in the short term. A decelerating universe would be a comfort because it offers the prospect of more to see. The limit of our vision in cosmology is the horizon, by analogy with the limit of a view of an Earthly landscape. The horizon in a decelerating universe grows with time, because light can reach us from more remote regions, and regions that were moving apart faster than the speed of light early on slow down, allowing light to "catch up." The observable universe gets bigger every day.

An accelerating universe carries distant regions away from us faster and faster. Anything not bound by gravity will be pulled from view. The deceleration is like a fisherman with lots of lines out and lots of fish on the ends of the lines. He willingly lets them play out because he knows the fish will tire and slow down and he'll pull them in and have a fine supper. The acceleration is like a fisherman who's dismayed to see all the fish race away faster and faster, giving him no chance to reel them in. He goes hungry.

Michael Busha and his colleagues at the University of Michigan have worked out the consequences of the accelerating universe, assuming that dark energy doesn't vary across space and time.[10] Why does the universe accelerate? Because dark energy is an intrinsic property of the vacuum and as space expands there's more vacuum and so more dark energy, creating more space. It's a runaway effect that leads to an exponential expansion.

Gravity can't create new structures in the face of the acceleration and existing structures will be more and more isolated. Galaxy clusters and groups become tiny dimples in rapidly growing space-time, each one a miniature "island universe" wrenched from contact with other clusters or groups. The mass in these isolated structures is trillions of times smaller than the dark energy contained within the horizon. Matter becomes both lonely and insignificant.

Here are some of the landmarks of this scenario.[11] Over the next few billion years, future astronomers will lose sight of the distant, ancient universe as photons that were once approaching us are dragged away by faster than light expansion. In 120 billion years, when the universe is 1000 times bigger than it is today, the Virgo Cluster will move past our horizon and Milkomeda will be the only galaxy to look at. How sad the fate of cosmologists, reduced to staring at their own navels.

It gets even worse. The next landmarks happen after about a trillion years, a time at which the lowest-mass stars in galaxies like the Milky Way are still feebly shining. Radiation is stretched by the expansion, and the microwaves from creation, already stretched by a factor of 1000 between 380,000 years after the big bang and now, will have been stretched by another factor of 10^{28}. Those photons are as big as the horizon, and the universe no longer has a big enough "box" to hold them. A mere 140 billion years later, after another growth factor of 10,000, starlight leaving our galaxy or any other is stretched beyond the horizon. There will be no photons within these island universes, just their ghosts manifested as slowly varying electric fields.

Whoever might live in the universe after these events will be stuck in their own pocket of space-time, drained of matter and radiation, with an extremely limited view. Each horizon is like an event horizon, so it will be like being trapped in a black hole. Stephen Hawking theorizes that feeble radiation will leak into each horizon with a temperature of just 10^{-29} Kelvin ($-460°F$). An inhabitant might react as Galileo did to his fate under house arrest, blinded by too many careless observations of the Sun: "This universe that I have extended a thousand times . . . has now shrunk to the narrow confines of my own body."

The Big Rip

The possible fates of the universe used to be simple to describe and relatively easy to understand. There were two possibilities: fire or ice. When Robert Frost penned his poem about the fate of the universe he posed a dichotomy. Either there's insufficient matter to overcome the cosmic expansion, leading to a universe that grows larger, colder, and more rarified. Forever. Or the cumulative effect of everything tugging on everything else causes the universe to reach a maximum size, sigh (in a metaphorical sense), and collapse in a reverse chronology of the big bang. By the late 1970s, astronomers had a good enough census of the dark matter to conclude there wasn't enough to overcome the expansion. The universe would expand forever.

The conclusion is still the same: There will be no "big crunch" and the poetic elegance of hot or cold death is moot. But dark energy is a new and unfamiliar beast that changes everything. We know so little about it that theorists have given free rein to their fecund imaginations.

The simplest option is Einstein's cosmological constant—dark energy that's the same everywhere in the universe and doesn't evolve with time. We've already encountered the problem with this, the fact that dark energy is 120 orders of magnitude smaller than the value it would naturally have in current theories of particle physics. This was the motivation for quintessence, which ascribes acceleration to the effects of a massive hypothetical fundamental particle. Dark energy can be arranged to have its large natural value in the early universe and evolve to a very small value today.

The most extreme form of quintessence is phantom energy. In a 2003 paper, Robert Caldwell at Dartmouth College and his colleagues were disarmingly honest about their motivation for exploring it: "To be sure, phantom energy is not something any theorist would have expected; on the other hand, not many more theorists anticipated a cosmological constant!"[12] Phantom energy is the universe run amok. More than expansion, more than acceleration, it means the *acceleration* is increasing. The universe expands to an infinite size in a finite time.

Here's what would happen. The strength of dark energy would grow so dramatically that it not only would unbind any object bound by gravity, but it would also eventually exceed the other fundamental forces and unglue matter. About a billion years from now, phantom energy would rip clusters apart. It also probably would grab Andromeda before it could approach the Milky Way, depriving us of the pleasure of watching the formation of Milkomeda. Everything after that would occur in a crescendo that culminates 20 billion years from now in an event called the "big rip," so times are counted down from then (Figure 11.5).

Sixty million years before the big rip, the Milky Way would be torn asunder. Three months before the end of time, Earth and the other planets would be ripped from the Sun. Anyone standing on Earth as it flies apart would realize the dark energy has escalated to the strength of interatomic forces and the scale of planets; that signifies 30 minutes until the big rip. Atoms would be shredded a mere 10^{-19} seconds before the end, at which point the universe will have been completely disassembled.

A big rip is analogous enough to a big bang to be vaguely satisfying, but it's dire enough that it would be nice to show that it won't happen. The jury will probably be out until *Planck* returns the bulk of its data, but recent observations provide hints that the dark energy has the uniformity of Einstein's cosmological constant. With relief, we may be able to take the big rip off the worry list.

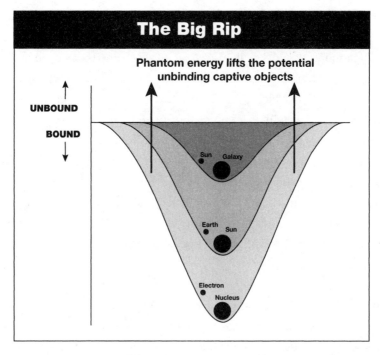

Figure 11.5. In the "big rip" model of cosmology, a particular form of dark energy acts to fuel the accelerated expansion of the universe. As a result, all the forces that act like "glue" would be overcome, first as stars are detached from galaxies, then as planets are detached from stars, and finally as atoms are broken apart.

Astronomers like the thrill of the chase and astronomer Adam Reiss, one of the co-discoverers of cosmic acceleration in the 1990s, put it this way, "If this was a fox hunt and dark energy was the fox, I think we have closed off another escape route. But there is still a lot of terrain left for the fox, and we've seen little more than a glimmer of fur."[13]

The Slow Fizz

William Bulter Yeats wrote at the beginning of the poem *The Second Coming*, "Things fall apart; the center cannot hold; mere anarchy is loosed upon the world." After the stars have died, after the galaxies have merged, and after space has stretched until it shimmers with dark energy and little else, the last redoubt of the universe is matter. We take its obdurate permanence for granted but favored theories of fundamental particles predict that normal matter will one day decay.

Not soon. Not in a trillion years. But in a time that's as long compared to a trillion years as a trillion years is to a billion billionth of a second. Radioactive elements decay with a wide range of timescales and free neutrons decay in 15 minutes, but in general, atoms seem eternal. Nobody has even seen a proton or an electron decay. To put a limit on the lifetime of a proton, you can either watch one for a very long time or a huge number for a shorter time and hope to catch the rare event. Physicists in Japan employ an experiment called Super-Kamiokande, a huge underground tank of ultrapure water, to look for proton decays. They haven't seen one yet, which puts a lower limit of 10^{35} years on the lifetime of a proton.

Why are they so sanguine that normal matter decays when they've never observed it? The early universe had a tiny excess of matter over antimatter, such that a split second after the big bang, when matter and antimatter annihilated, the residue became the 10^{80} particles and much larger number of photons of the present-day universe. The slight asymmetry between matter and antimatter can be explained by grand unified theories of the forces of nature, and proton decay is predicted by those theories.

Our guide once more is Fred Adams, the guru of deep time. His pouchy eyes allude to the weariness of waiting for normal matter to decay. In the far future universe, most objects are slowly cooling embers: white dwarfs. Protons decay into positrons and pions and then the positrons annihilate with electrons to produce gamma rays and the pions decay into gamma rays. This feeble flux of photons diffuses out through the white dwarf, adding about 400 watts to its output. An object the size of Earth with the power of three lightbulbs and a temperature of one-tenth a degree above absolute zero isn't much of a beacon. All the white dwarfs in Milkomeda won't equal the power of the Sun in its prime.

In this late phase of evolution, stars find their own fountain of youth. Proton decay lowers the atomic number of nuclei in the white dwarfs as well as lowering the overall mass. The star loses 90 percent of its mass and its carbon turns into helium and then hydrogen, returning it to the simplicity of the material it started with.

The same fate awaits neutron stars. Planets in fact get there first due to their lower mass. The rocky mantle and metal core of Earth will be turned into hydrogen in about 10^{38} years. White dwarfs don't meet this fate until 10^{39} years have passed. At this point, the universe no longer contains stars. Anyone hoping to warm his hands by a fire of any kind will have to turn to those ciphers of gravity, black holes.

In 1974 Stephen Hawking predicted that black holes are not entirely black. Due to quantum mechanical effects they have a temperature and emit a tiny amount of radiation. Stellar mass black holes radiate just 10^{-28} watts, a pathetic amount compared to the feeble energy of white dwarfs undergoing proton decay. They take a phenomenal 10^{65} years to evaporate. Even if you let that number bounce around in your head for a while it's unlikely to be meaningful. As the mass decreases the evaporation rate and energy released increase so the far future will see a wave of fizz as massive stellar remnants disappear (Figure 11.6). From the mass of a freight car into nothing takes just a second, and each one releases gamma rays equal to 5000 trillion metric tons of TNT or 1000 times the world's nuclear arsenals.

More massive black holes are rare, and evaporate slower, so they will be the last to go. The million-solar-mass black holes at the center of small galaxies take 10^{83} years to disappear and the billion-solar-mass beast that's

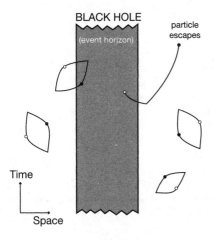

Figure 11.6. Black holes are not black. According to a theory of Stephen Hawking's, when particle/antiparticle pairs are created near the event horizon, one member of the pair can escape, creating other particles and photons. As a result, black holes slowly evaporate, at a rate that increases with decreasing mass.

likely to be left at the center of Milkomeda will take 10^{98} years, ending in the same crescendo of gamma rays as the smaller black holes.[14]

If 10^{98} years isn't infinity, you can see it from there. What more could there be to talk about? In the realm beyond 10^{100} years, protons have decayed, stars have dissipated, and black holes have evaporated. The waste products are neutrinos, electrons, and positrons, and photons of wavelength larger than the visible universe. All physical processes at this time, and even before, are subject to the uncertain nature of dark matter and dark energy.

At this point, Fred Adams isn't the only person to have bags under his eyes. We too, the reader and the writer, may feel like we've pulled an all-nighter. Our patch of the accelerating universe has disintegrated under cover of darkness into a uniform soup of particles and photons. The assembly and disassembly of stars and galaxies is interesting but it's like rearranging chairs on the deck of the *Titanic*. Entropy, calling card of the second law of thermodynamics, is the implacable winner. Regrettably, uniform chaos is the enemy of creativity and life. It was fun while it lasted.

Chapter 12

BEYOND ENDINGS

Nick Bostrom sends a precisely calibrated mixed message. His street cred as an analytic philosopher is beyond reproach, so his audience takes what he says very seriously. Older scholars listen intently; the young ones take copious notes. But the twinkle in his eyes and the occasional flash of a playful smile show that he recognizes the absurdity of his topic.

Bostrom is presenting what he sees as the logical consequence of the likelihood that the universe contains many creatures whose intelligence and technology dwarfs ours. With vast computational power at their disposal, these cultures could easily replicate the entire history of thought processes of the human race. And that—now he pauses a beat to make sure he has the audience's full attention—means we're almost certainly living in a computer simulation.

Bostrom is one of a new generation of philosophers who've absorbed the amazing information of modern astronomy. With biology very likely to be universal, and trillions of potential sites for life in the universe, they ask what truly advanced sentient organisms might do. With genesis as a space-time foam, they speculate about hypothetical and parallel universes. And, being philosophers, they're bound to ask deep questions about the meaning of it all and just how real this thing we call reality is.

Twenty-five centuries ago, Plato told the allegory of cave dwellers who yearned to understand nature yet who were transfixed by its imperfect projection as shadows on the cave wall. Scientists like to think they deal in reality. Are they just chasing shadows?

Although he gained notoriety and media attention for his simulation hypothesis, Bostrom's best and most detailed work involves the "anthropic" principle. Scientists have noticed that a number of physical properties of the universe, from the strength of the fundamental forces to the expansion rate itself, are set at values that permit the existence of carbon, water, stars, and life. If these properties were slightly different, the universe would still exist but it wouldn't be alive.

Was the universe built for life? Rather than jump to that astonishing conclusion, Bostrom has subjected anthropic ideas to his brand of formal and logical rigor. He finds that a self-selection principle is operating; we must be wary of ascribing special significance to features of the universe that are necessary for our existence. If the universe could speak it may be saying to humans: Get over yourselves.

Living in the Multiverse

Fine-Tuning

A physicist runs up to you in great excitement. He pauses to catch his breath and the words tumble out. It's amazing, he says, if the strong force that holds atomic nuclei were just a bit stronger, stars would turn all their hydrogen quickly into helium and on up to iron. If it were a bit weaker, there would be no complex nuclei formed and no carbon. And if the weak nuclear force were a bit stronger, neutrons would decay too fast for stars to build heavy elements, and if it were a little weaker, all the hydrogen would be used up.

He stops for a minute just to catch his breath and you try to politely edge away, but he's not done. If the electromagnetic force were a bit different in strength, molecules would not form so chemistry would be impossible. And think about gravity, he says. If it were much stronger, stars would live quick lives, taking millions and not billions of years. If it were much weaker, stars could not form heavy elements. He grips your arm, a little too fiercely. Don't you see, he says insistently, in all these cases, it would be a universe with sensible physical laws, but no life would be possible. And he stands, hands on hips, waiting for your reaction.

Are you surprised? Should you be? These questions are at the heart of a lively debate over fine-tuning in physics. Standard physics has only a few dozen parameters and many of them would have to be very close to their observed values for biology and creatures like us to exist.[1]

Fine-tuning arguments extend to the realm of cosmology. Physically plausible universes could have much more or much less matter than our universe. But if the universe had much less matter it would have expanded so rapidly in the early phase that no stars or galaxies could have congealed from the cooling gas. A structureless universe would almost certainly be devoid of life. On the other hand, if the universe had much more matter, it would have recollapsed long ago, without giving

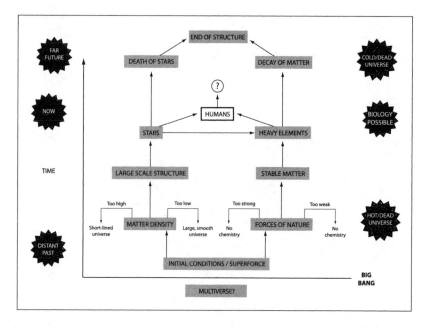

Figure 12.1. Fine-tuning arguments suppose an important connection between the substrate reality of forces of nature and the mathematical theory that accounts for them, and particular features of the nearby universe that lead to stars and planets and biology. The features are "special" because they lead to us but that doesn't mean they have to be preordained or built into the initial conditions.

stars the chance to make heavy elements and shelter biology. Perhaps only our type of large and long-lived universe can support observers like us (Figure 12.1).

It seems striking. These "counterfactual" universes aren't ridiculous; they're physically sensible. The biological counterfactual of imagining pigs that could fly isn't very edifying because we would have to make a tortuous reasoning based on evolution and environment to see how pigs could evolve wings and the ability to fly. It's not impossible, but it's not very interesting. In the cosmology situation we haven't thrown away core physics ideas like causality or the conversion of matter into energy—we've simply tweaked the forces a bit and found that we get universes where life as we know it is essentially impossible.

But for fine-tuning to be surprising we would have to show that the values of the physical constants being within a small range of their actual values is an unlikely event. There's an assumption in the logic that physical quantities can take a huge range of hypothetical values and the probability is more or less uniform across that range. But the assumption is pulled out of thin air.

It's not quite the same as fine-tuning, but our place in the universe doesn't support the Copernican principle or the mediocrity principle. We do orbit a normal star in a normal galaxy, but in other ways our situation seems special. Most entities in the universe are photons or dark-matter particles but we're composed of protons and neutrons. Most of the universe is an almost perfect vacuum but we live in a galaxy. Most of the Solar System is contained in a star yet we live on a rocky planet. Finally, the lifetime of the universe is enormous, yet we find ourselves living within a slender interval eons after the big bang. Is this also surprising?

The Anthropic Principle

How can we possibly explain the fine-tuning of the universe around values of physical quantities that permit biology, when a seemingly slight deviation from those values would render the universe sterile? To put it another way, what we observe is a small subset of possible realities, but most counterfactual realities exclude us (Figure 12.2).

At a conference in 1973 to celebrate the 500th anniversary of the birth of Nicklaus Copernicus, astrophysicist Brandon Carter coined the term "anthropic principle" to refer to the anti-Copernican stance that we *do* live in a privileged situation in space and time. He framed a weak and a strong form of the principle. The idea connects to fine-tuning but is even more controversial, in part because it gives intelligent observers a central role in the universe and in part because there have been so many definitions in the research literature.

The weak anthropic principle says that we can only observe a universe that has properties such that intelligent observers can exist. People are inclined to say nothing and blink when they hear this—it seems like a truism or a tautology. The strong form says that the universe (and the physical parameters on which it depends) must have values that admit the creation and existence of intelligent observers.

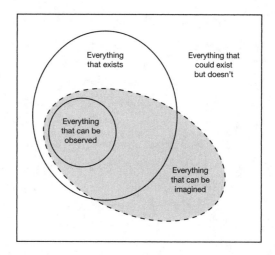

Figure 12.2. We observe a small fraction of the physical universe that is postulated in our theory of cosmology. That in turn is a small fraction of possible realties, most of which would be inhospitable to life. Meanwhile, the imagination (shaded region) includes things that exist but haven't been observed and things that may not actually exist.

Playwright and novelist Michael Frayn put it this way in his book *The Human Touch*: "It's this simple paradox. The universe is very old and very large. Humankind, by comparison, is only a tiny disturbance in one small corner of it—and a very recent one. Yet the universe is only very large and very old because we are here to say it is. . . . And yet, of course, we all know perfectly well that it is what it is whether we are here or not."[2]

There are even more extreme or audacious versions. A "participatory" form of the principle is motivated by physicist John Wheeler's idea that quantum effects create a deep connection between observers and the larger space container. In effect, observers create the universe. John Barrow and Frank Tipler's book *The Cosmological Anthropic Principle* ends in a flight of fancy regarding the Omega Point, when complexity and intelligence evolve to their limit: "Life will have gained control of *all* matter and forces not only in a single universe, but in all universes whose existence is logically possible; life will have spread to *all* spatial regions in all universes which could logically exist, and will have stored an infinite amount of information, including *all* bits of knowledge which it is logically possible to know. And this is the end."[3]

When the balloon of an idea gets so big and stretches so thin, there are people happy to prick the bubble. The strong form of the principle is the most tempting to smack down because it smacks of teleology, the idea that the universe has a purpose. The strong anthropic principle has been co-opted by supporters of creationism and intelligent design in arguments not much different from those penned by mathematician and philosopher Gottfried Leibniz 300 years ago: "Now, as there is an infinity of possible universes in the Ideas of God, and as only one of them can exist, there must be a sufficient reason for God's choice, which determines him toward one rather than another."

Another critique comes from logic. Just as we shouldn't be surprised that the universe has conditions consistent with our existence, we shouldn't be surprised that it doesn't have conditions that would be inconsistent with our existence, even though those conditions might not violate certain physical theories. The anthropic principle seems to invert cause and effect. Harvard paleontologist Stephen Jay Gould once compared the claim that the universe is fine-tuned for the benefit of our kind of life to saying that sausages were made long and narrow so that they could fit into hotdog buns, or that ships had been invented to house barnacles.

Then there's the central role given to intelligent observers. What's so special about intelligent observers and why should we be the template for them? There's no generalized theory of biology, and certainly not a theory of the pathways to intelligence, that requires an essential role for carbon or long-lived stars. If life exists in a wider range of physical parameters than we imagine from our single example, fine-tuning may not be as severe as we think.

Nobel Prize–winning physicist Steven Weinberg has worked on anthropic ideas but knows they're slightly disreputable. He says "A physicist talking about the anthropic principle runs the same risk as a cleric talking about pornography: No matter how much you say you're against it, some people will think you are a little too interested."[4] Despite the criticisms the anthropic principle has received lots of attention and advocacy from esteemed scholars like Weinberg. This is because it meshes neatly with frontier ideas in cosmology that try to speculate meaningfully about space-time beyond our horizon.

Welcome to the Multiverse

Leave your intuition at the door as we explore "the landscape." String theory is an ultimate theory of nature that aims to unify all forces and all particles in a mathematical framework that incorporates quantum mechanics and general relativity. In most formulations, string theory has a mind-bending 11 dimensions, but a universe with 3 dimensions of space and 1 of time is easily accommodated. The general objects in the theory are branes (short for "membrane"), and they dynamically occupy space-time in an arbitrary number of dimensions: 0-branes are points, 1-branes are strings, 2-branes are sheets, and on up to higher dimensions. String theory would move from an exotic speculation to a mainstream idea if we could somehow test for the "extra" dimensions; currently that's not possible (Figure 12.3).

String theory meets cosmology in the vacuum. Nothing might be the most interesting thing of all. The theory predicts an incredible number of vacuum states with randomly different properties: roughly

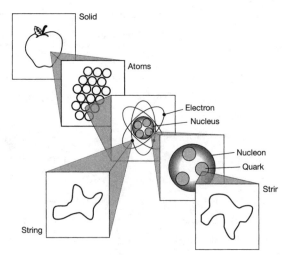

Figure 12.3. String theory posits a deeper level of reality based on 10-dimensional space-time. In the conventional theory of matter, quarks and electrons are fundamental. In string theory, they and all other particles are composed of one-dimensional strings. The hidden dimensions are only manifested at extremely high energies.

10^{500}, according to experts. That's the "landscape." Remember that in modern physics no vacuum is ever empty—it contains electromagnetic waves and particles and antiparticles popping in and out of existence all the time. Our universe resulted from the spontaneous inflation of one of these quantum events. Inflation stretched this quantum seed to a size far bigger than we can observe, so the universe appears uniform and smooth and the laws of physics are the same everywhere within it.

According to Martin Rees, the string theory landscape provides a basis for a prodigious, and perhaps infinite, ensemble of universes called the multiverse. The other universes might be large or small, ephemeral or eternal, living or dead, exciting or humdrum. Rees is the eminence grise of cosmology, England's Astronomer Royal. Lord Rees of Ludlow, as he's formally known, jokes that one of his jobs is to cast horoscopes for the Queen. He's a small man with gentle eyes and the cadence of a preacher from southern England. His reassuring tone of voice makes outlandish ideas seem comforting and reasonable. He writes, "This new concept is, potentially, as drastic an enlargement of our cosmic perspective as the shift from pre-Copernican ideas to the realization that the Earth is orbiting a typical star on the edge of the Milky Way."[5]

The immense number of physical states of the multiverse neutralizes the fine-tuning argument. Yes, physical law adheres closely to what seems to be required by life and the dark energy is anomalously small by a large factor. But with 10^{500} vacuum states, it's not unlikely that the universe has the properties we see. We naturally live in one of the habitable universes; the much more abundant uninhabitable ones are elsewhere in the string theory landscape, unobservable by us.

The landscape might be populated by different regions of space, as in Stanford physicist Andrei Linde's chaotic inflation theory, where space-time bubbles appear and disappear endlessly. Or they might be different eras of time in a single big bang. Or they might be bubbles of space-time that form within bubbles, in a progression without end. Some people have criticized the multiverse theory for not being predictive enough. Steven Weinberg disagrees, saying, "The test of a physical theory is not that *everything* in it is observable and *every* prediction it makes is testable, but rather that enough really is observable and enough predictions are testable to give us confidence that the theory is right."[6] The jury is still out on the multiverse, but a few people are pretty confident.

At a symposium on the multiverse at Stanford University in 2003 Rees gave a public talk along with Paul Davies from Arizona State University and Andrei Linde from Stanford. At the very end, an audience member asked how much each of them would bet that the multiverse theory was true. Rees recalled a physicist who was asked whether he would stake his goldfish, dog, or child that his theory was correct, and replied, "I would stake my dog that there's a multiverse out there."[7] Davies agreed, "Me too, at the dog level. The top level is all mathematical." The last to answer was Linde, who got a big laugh from the audience by saying, "I'd bet my life!" Two years later at a conference in Cambridge, Steven Weinberg upped the ante: "I have enough confidence in the multiverse to bet the lives of both Andrei Linde and Martin Rees's dog."

From Endings to Meaning

Endless Twilight

The anthropic principle has left the door ajar on the intriguing idea that the universe is built around life. In the context of string theory and the multiverse there's an unimaginably large real estate of space-time and physical variation and we occupy one of the rare universes that are "interesting" because complexity has resulted in biology and sentience.

What makes this universe noteworthy is the fact that it has spawned (at least) one type of creature who can reflect on its existence in the construct. The many other universes have stories to tell but probably no narrators; they are of purely academic interest, like the rocks in a mineral museum.

Let's explore the ultimate limitations on life in the universe, starting with the patch of space-time we inhabit then extending to the totality of space-time in the multiverse. In 1979, physicist Freeman Dyson wrote an influential paper called "Time Without End: Physics and Biology in an Open Universe."[8] Dyson is a visionary who has written extensively about interstellar travel and space colonies and he was influenced by Olaf Stapledon's *Star Maker* and other science fiction growing up. He's also a staunch proponent of nuclear disarmament. Dyson admits mixed success as a prophet but says, "It's better to be wrong than to be vague."

Dyson analyzed the dark and cold future of an endlessly expanding universe and reached a surprisingly sanguine conclusion. He defines the thermodynamic measure of information contained in a complex organism, and finds it to be 10^{23} bits, or 10^{33} bits for all humans on the planet. We are profligate with energy; we waste it and radiate it into space as if it were an endless resource. But in the distant future, when stars are all dead and space heads toward absolute cold, life will have to be miserly with energy. This could happen in two ways. The first is by matching the metabolism to the falling temperature so that energy from the feeble source can be radiated away. The other is by hibernating in increasingly long cycles. We sleep a third of our lives and don't begrudge that; creatures of the far future will be awake for smaller and smaller slivers of the future eons.

Given that strategy, Dyson showed that energy equivalent to a mere eight hours of sunlight could sustain a human civilization forever. He thinks future life will not be "wet" biology but will be interstellar gas clouds. This was a conceit used by Sir Fred Hoyle in his 1957 science fiction novel *The Black Cloud*, but Dyson takes it absolutely seriously: "An ever-expanding network of charged dust particles, communicating by electromagnetic forces, has all the complexity needed for thinking an infinite number of novel thoughts."

What would such exotic entities do with their increasingly rare waking moments? Recreation is difficult and pointless in such a cold and dark universe. Sex is probably out of the question for a gas cloud. That just leaves pure thought. They could work through all the possible moves in a game of chess at a trillion years per move and still have time left over. Our restless monkey brains make patience a rare virtue; in the far future it will be a necessity.

For a while, this almost rosy scenario was in doubt. Dyson got into a technical and esoteric debate with two physicists who claimed that his recipe for infinite life didn't work. Dyson was pleased by the challenge, saying, "It's much more fun to be contradicted than to be ignored." The argument was that any material system that stores information has a fixed minimum energy state with a gap above it to the next energy state. Once the temperature gets too small, the system can't absorb and emit energy so it can't store information. Dyson accepted this but said that it only applied to digital life. If life is analog, as in his favored dark cloud, or as in familiar biology, it can persist by growing in size.

The fate of our patch of space-time seems clear if dark energy keeps dictating the expansion, but what about the larger construct? Inflation gives a framework for hypothesizing the multiverse but no predictions that could test the idea. Aspects of inflation involve fine-tuning and the strength of dark energy is a mystery in the model. As a result theorists have continued to search for an elegant way to explain the universe.

In 2001, physicists Paul Steinhardt from Princeton and Neil Turok from Cambridge caused a stir with their cyclic universe theory. It echoes ancient mythology and philosophy by saying that time and space exist forever. The universe goes through endless cycles of expansion and contraction with trillions of years of evolution in between. They avoid the problems of the initial singularity—with its infinite temperature and density—and they don't need inflation because the flatness and smoothness of the universe were caused by events that preceded the most recent big bang.[9]

Steinhardt and Turok use a form of string theory called "M-theory" in which the universe is a three-dimensional brane that exists and moves with other branes in a higher dimensional space. When branes collide they create a burst of energy that we experience as the big bang and an expanding universe. After dissipation and contraction, the two branes eventually come into contact again and trigger a new big bang and a new expansion cycle. Cosmology is a mature science, so cyclic theory has already had to explain an impressive number of observations just to be taken seriously. The critical testing ground will be gravity waves. The standard big bang predicts that inflation flooded space with gravity waves, while colliding branes produce no gravity waves. The decisive observations will come from *Planck* and upcoming gravity wave experiments.

Room at the Bottom

It's hard to imagine what intelligent species remote from us in either space or time might do with their capabilities. We are, after all, very young and immature. In the futurist parlance, we're far short of being even a type 1 civilization, since we use only a millionth of the energy that falls on our planet. Call us type 0; we're barely on the radar of sophistication. The next jump is a factor of a billion to harness all the energy of a star, which is a type 2 civilization. Beyond that, a type 3 civilization can harness the energy of an entire galaxy. Surely species like that would treat space-time as a plaything, navigating wormholes and creating new universes the way we drive cars and build houses.

If that's too fanciful, consider this analogy. We're just a billion years removed from single-celled organisms. Complex life in other parts of the universe could have had a 10-billion-year head start on us. They might be to us as we are to anaerobic bacteria—tenfold.

With all the potential paths for life in the universe, it's worth looking beyond the "large and slow" strategy outlined by Dyson. Analogies between biological components and computers are flawed, but they give a sense of whether or not there's "room at the bottom" in terms of life that's small and fast and fits more into available cosmic time.

Our cells pack a lot of information into a 1/100 millimeter package: a gigabyte is stored in the four-letter base pair alphabet of the genome and several gigabytes more of memory in the shape and location of millions of biomolecules that carry out cell functions. That's equal to one DVD's worth in a container the size of a dust mote, or 10^{16} bits per cubic millimeter. The brain has 10^{11} neurons and 10^{14} synaptic connections in a 1200-cubic-centimeter package. Carnegie-Mellon roboticist Hans Moravec puts human processing power at 100 trillion calculations per second and memory at 10^{15} bits. That's only a million bits per cubic millimeter. However, the brain employs a massive parallelism that eclipses our best supercomputers.

Biology on Earth might be optimized by billions of years of evolution. Perhaps competition and environmental drivers elsewhere could lead to greater speed and processing power but that's pure speculation. If creatures here or on another planet pass through what Ray Kurzweil called the singularity and become postbiological, the gains could be fantastic. The continuation of current solid-state technology can give speed gains of a factor of a million over neutrons, with another factor of 10,000 possible before the limit of silicon is reached. MIT professor Seth Lloyd, who refers to himself as a "quantum mechanic," has written about the ultimate physical limits to computation.[10] An optimized, one-kilogram computer made of normal matter, which Lloyd calls the "ultimate" laptop, could process information at 10^{41} operations per second and store 10^{31} bits.

The gain in information storage density over biology is a factor of a billion, which comes from storing information in the energy states of single atoms as opposed to storing information in molecule shapes. However, the gain in speed is a phenomenal factor of 10^{27}! If we or any other species is ever able to transfer brain functions into silicon we'll be able to realize this potential. If doing your taxes took 10^{-21} seconds instead of a week think of how it would liberate you to have greater thoughts. One way to counter the accelerating universe is to accelerate our brains.

We have implicitly abstracted life into computation. The fundamental limit to computation is set by the speed of energy changes in quantum mechanics. Going beyond normal matter, the most efficient storage of information occurs inside a black hole, which is called the Beckenstein bound. A black hole of the same mass as the ultimate laptop described above would operate 10 billion times faster. This all sounds like crazy science fiction until you realize we'll get there in another 200 years of Moore's law.

If you thought a black hole was the pit of despair from which nothing can escape, there's hope. Black holes embody an information paradox because no matter what went into them—socks, people, encyclopedias—they look the same (Figure 12.4). The radiation predicted by Hawking is part of the problem; it has no structure, so information

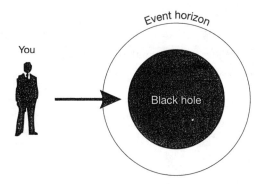

Much, much later...

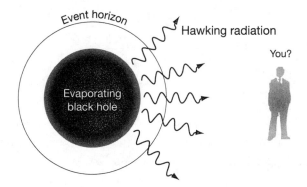

Figure 12.4. Black holes embody an information paradox, because anything falling in only adds to their mass and all its exact information or structure is later turned into pure radiation. But if the information does come out as Hawking radiation, it's not clear how that unstructured radiation can ever truly represent the person who was lost.

seems to be lost. In 1997, physicist John Preskill at Caltech bet Hawking that he was wrong and in 2005 Hawking conceded the bet after deciding that quantum effects near the event horizon could allow information to get out. It turns out black holes may store and process information.

Is this an escape clause? Black holes have long been suspected as a path to immortality because, seen from afar, time slows down without limit as the event horizon is approached. You would need to find one at least a million times the Sun's mass; any smaller and tidal forces would rip you apart as you fell in. To people watching, you'd slow down as you approached the horizon, until you reached a frozen moment, forever young. It's a Pyrrhic victory because your experience is to fall into the black hole toward an uncertain future. It's unknown whether you would be disassembled and regurgitated into Hawking radiation, deposited via a white hole to another part of this or another universe, or obliterated at the singularity. Think of it as a last resort.

Meet the Sims

Before considering astronomical and unlikely outcomes, we covered the odds of dying in mundane but likely ways, by disease or accident. Some scholars have placed similarly high odds on the hypothesis that we live in a computer simulation.

Nick Bostrom is the most prominent purveyor of this idea so it's worth keeping in mind his unsettling mixture of ironic Swedish detachment and deadly seriousness. He directs a prominent Oxford think tank yet has dabbled as a stand-up comedian in London clubs. Bostrom starts by assuming substrate independence, the idea that if all the physical attributes of a brain were rendered computationally, the result would be consciousness. He then estimates the number of computer operations required to emulate the history of all thought processes of everyone who has ever lived. That number has to include the environment, but only with appropriate verisimilitude—our imme-diate surroundings and everyday objects in detail, but the night sky only at the level of detail of the largest telescopes, and Earth's inte-rior and microscopic details omitted. His conservative estimate is 10^{36} operations, and even if it's off by a few orders of magnitude the argu-ment is unaffected.

It's called an ancestor simulation and it's not that hard to do. We'll be able to run them in another 50 years of Moore's law and Seth Lloyd's ultimate laptop could churn out 100,000 simulated civilizations every second. Unless we're the first civilization to approach and attain this level of technological sophistication, other civilizations have already created simulated creatures like (or unlike) us. Because it takes such a small fraction of their resources to run huge numbers of simulations, few civilizations have to get to that point before simulated ancestors greatly outnumber real ones. By the Copernican principle we're very likely to be simulated.

Welcome to *The Matrix*, but without the part when real human bodies are suspended in vats to act as batteries. In the simulation hypothesis, you have no body; your body and brain are the creation of somebody else's computer program. There's a reflexive revulsion to the idea. Of course I'm real! People will naturally think that way, but they're all wrong. There would only be "glitches" in the simulation or hints of the truth if the simulators so desired.

Logically there are just two alternatives to the conclusion that we live in a simulation. One is that civilizations self-destruct before getting to the point where they can run ancestor simulations. That's bad news, because it would be a likely outcome for us too. It has to be a strong bottleneck; the few getting through it could flood the world with sims. The other is that advanced civilizations choose not to run simulations. Perhaps they have moral qualms or are too advanced for such games. That would require unusual convergence in the motivations of diverse intelligent beings. It's counter to our experience because we increasingly re-create in the computer-created worlds (Figure 12.5).

Living in a simulation resolves some long-standing questions. Evil is not the result of some paradoxically flawed Creator; it was added by the simulators to liven up the game. The afterlife and supernatural experiences are "features" and don't have to be explained away by skeptical scientists. Free will is a convenient and potent illusion, and fear of death keeps people engaged in the simulation. Even if you're convinced you're simulated the accuracy of the construct means you have no strong reason to change your behavior or worldview. If we are simulated it's plausible that our creators are simulated too, and their creators in turn might be simulated, and so on. Nobody knows how many levels of "reality" there could be.

Figure 12.5. The author's avatar, Cosmo Priestman, strolling in a virtual "Garden of Eden" called Svarga in the three-dimensional virtual world Second Life. Visual realism is coupled to the ability to communicate by voice with other avatars. Consider this a zero*th*-order simulation, then try to imagine what might be possible with hundreds or thousands of years more development.

Solipsist Sonata

With the simulation hypothesis, we teeter on the edge of solipsism. If nothing is real, or if only our own thoughts are real, we're sealed off from the universe. Which would be a waste because awareness of the universe of which we're a tiny part is one of our greatest triumphs.

The story of cosmic endings is science at its best: asking the biggest questions, all cylinders firing, bending theory and brain power to the task. It's also science at its most uncertain. If physics could be used to predict the future state of the cosmos and everything in it, our free will would be the biggest illusion of all (Figure 12.6). But there's no way to poke and prod the universe as we can something in the lab and most of it isn't even visible to us. As the Russian physicist Lev Landau wryly noted, "Cosmologists are often in error but never in doubt." This last chapter implicitly bears the inscription that some feel should be on the door of all churches: Important if true.

I'm an astronomer. I routinely observe things that dwarf me in space and time, but I probably keep the universe at a healthy remove. After all, tightrope walkers don't dwell on their frailty in the face of gravity. In one unlikely moment several years ago, I became aware of the size of the universe and my finiteness in space and time. I was on holiday in the Caribbean, lying on my back in shallow water, finally relaxed in the third day of a two-week vacation. The Sun was setting behind my head, the water was bathwater warm, and small waves slapped at me playfully.

As the Sun slid below the horizon like a gold coin slipping into a back pocket I had a Copernican realization of the Earth's rotation pivoting me toward the emerging stars and the blackness of space. Spread-eagled, I felt the planet curve away from me in every direction. The exposure was shocking. Nearly naked and in womblike water, I was thrust into the cosmos and my awareness quickly passed through the familiar confines of the Solar System, out among the star fields and glowing gas clouds of the Galaxy, on into deep space, past pinwheel galaxies and the filigree of the cosmic web, to a shimmering paleness at the limit of vision. Awe paralyzed me, but almost immediately the moment passed and I receded into my familiar container, with the universe ensconced comfortably in my head. A full Moon rose like a milky eye between my feet.

Our personal encounter with the universe will end, but let's say we're persuaded by the argument that sentience isn't unique to this

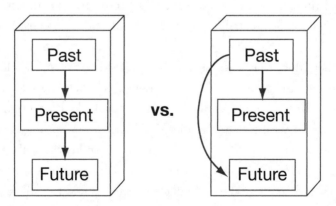

Figure 12.6. The present arrives from the past and heads into an uncertain future (*left*). If the most ambitious agenda of science were possible, we could predict the future from our measurements of the past (*right*). But locally real phenomena can't be used to infer chains of causality beyond our local space and time. Only if we could predict the exact state of the current universe from the initial state could we predict the future state with certainty.

planet. Let's imagine that humans have more than an academic interest in the progress and fate of sentience elsewhere—there's kinship in the fever that we call life.

The Drake equation was formulated to apply to the Milky Way but the visible universe contains 50 billion galaxies, so the cosmic estimate is $N = 10^{11}L$. The universe is likely to be teeming with intelligent life. If the technological phase is fleeting for most of them, that still leaves a tail of long-lived civilizations for whom the distance between galaxies can be accommodated by patience. To us, they are immortals.

Suppose we keep a lid on the nukes and duck the green goo and the gray goo and deflect the meteors and the comets. Suppose we don't solve the problem of senescence but survive to a venerable age as a species and send shoots and seedlings to nearby stars. Elsewhere our biological and postbiological "cousins" slowly percolate around this galaxy and billions of others, the rare encounters leading mostly to blank incomprehension, leavened by flashes of insight. Some life is curious and acquisitive, some is introspective, some is ethereal, and some is lethal.

The dying of the light culls all but the most persistent and ingenious. As the universe recedes from view, black hole engineers give up and abandon their "wormholes to nowhere," and the survivors huddle near the feeble evaporative glow of the central black hole in Milkomeda and tell timeless stories about time. It's been a trillion trillion trillion trillion trillion trillion trillion trillion years since the big bang. Everyone agrees: Life had a good run.

Where does that leave us, a flawed but exuberant species learning how to live up to our potential? Sentience is a blessing and a curse. Perhaps we're just pond scum that got a little lucky on a Friday night and woke up with a Saturday morning hangover of cosmic awareness and existential angst. Would it be better to be industrious like the ant or ephemeral like the mayfly, or be able to apply our brains to a finite world that's like a liquid womb, like the octopus and the whale? We're justified in our surprise that we're more than mindless matter. In this universe full of magical moments, it doesn't matter what happens in

THE END

Glossary

ACTUARY. A professional who deals with the financial implications of risk and uncertainty, including human death. Rated by *The Wall Street Journal* as the best job in the U.S. in 2002.

ALLELE. A member of a pair or a series of different forms of a gene. A single gene controls flower color, but different alleles will result in different colors.

ANDROMEDA. Also called M31, this spiral galaxy is 2.2 million light-years away and is similar to the Milky Way in size.

ANTHROPIC PRINCIPLE. The idea that the features of the universe are constructed around the existence of intelligent observers or life in general. A controversial philosophical premise.

ARCHAEA. The most recently discovered domain of microbial life and the one closest to the root of the tree of life.

BACKGROUND EXTINCTION. Average rate of disappearance of species, which can be difficult to measure reliably for rare or short-lived species, or species poorly sampled in the fossil record.

BECKENSTEIN BOUND. The ultimate limit on the density of storage of information, set by the theoretical properties of black holes.

BIG BANG. The hot, dense creation of the universe 13.7 billion years ago, supported by observations of galaxy redshifts, microwave background radiation, and the abundance of helium and other light elements.

BIG RIP. Hypothetical ending of the universe in some models of dark energy, where space expands exponentially and dark energy overcomes forces on smaller and smaller scales, eventually disrupting all matter.

BIODIVERSITY. The variation of life within an ecosystem or the entire planet, often used as a measure of biological health.

BIOMARKER. Signature of life on a remote planet or moon based on information derived from images or spectra.

BIOSPHERE. The total "system" of life on Earth. Most is on the land and in the air and oceans but a substantial fraction is under the surface; some parts may even be unrecognized.

BLACK HOLE. Remnant after a massive star dies at the end of its life, when all nuclear fuel has been exhausted. No matter or radiation of any kind can escape from a black hole.

BROWN DWARF. An object too small to get hot enough for any kind of fusion reaction, so it glows with infrared radiation.

CARBON CYCLE. An interlocking set of transactions of organic material among rocks, the oceans, the atmosphere, and life.

CENTENARIAN. Someone who has lived to the age of 100 or more. Only 1 in 5000 Americans is a centenarian.

CLIMATE CHANGE. General term for the long-term variations in global climate; causes can be geological or astronomical as well as the more recent changes caused by human activity.

CLONAL COLONY. A group of genetically identical individuals in one location that originated vegetatively, not sexually, from one ancestor. Can be plants, fungi, or bacteria.

COMPUTATIONAL ASTROPHYSICS. The discipline of simulating various important aspects of the universe in a computer.

COPERNICAN ARGUMENT. Any argument referenced to the insight of Copernicus that the Earth is not the center of the universe. In a general sense, the idea that Earth's position in time and space, or even humanity's current situation, is not special or unusual.

COUNTERFACTUAL. A state of reality or an outcome of evolution of the universe that is not observed, but could be in principle.

CRYONICS. Low-temperature preservation of people or animals, in the hope of future resuscitation. Not known to be reversible.

CYBORG. A human-robot hybrid, the hypothetical merger of humans with assistive technology.

CYCLIC UNIVERSE. The theory that our expanding universe is a phase in an endless series of universes, triggered by collisions of higher-dimension entities called branes.

DARK ENERGY. Attribute of the vacuum of space that's responsible for the acceleration of the universe. It's physical nature is unknown but it constitutes nearly 70 percent of the universe.

DARK MATTER. Enigmatic form of matter that dominates normal matter (protons, neutrons, and electrons) by a factor of 7 and is responsible for the gravity that holds galaxies together.

DRAKE EQUATION. A series of numerical factors that provides a rough estimate of the number of intelligent, communicable civilizations in the Milky Way galaxy.

EARTH-CROSSING ASTEROID. Asteroid with the potential to hit the Earth, although for most the possibility is extremely small. Also called near-Earth objects (or NEOs).

ENTROPY. A measure of the disorder of a system, with a tendency to increase over time. Biology reduces entropy in a localized way, because cells are highly structured, but at the expense of increasing the entropy of the external environment of a cell or organism.

EUKARYOTE. Cell where the genetic material is contained in a nucleus. The lineage that evolved to include all plants and animals.

EVAPORATION. The process in a galaxy where most stars are flung out from the galaxy while others fall into its core.

EVO-DEVO. Evolutionary developmental biology, the field that studies how developmental processes and environmental influences can lead to new features in an organism.

EXOPLANET. Also called extrasolar planet, this is a planet beyond the Solar System. The first was discovered in 1995; now nearly 400 are known. Most are similar to Jupiter in mass but the smallest approach the mass of Earth.

EXTREMOPHILE. Organism adapted to physical conditions beyond the normal range of temperature, pressure, pH, salinity, hydration, or radiation of most animals. Most, but not all, are microbial.

FERMI QUESTION. The question "Where are they?" as posed by the physicist Enrico Fermi, in reference to the surprising absence of evidence of any extraterrestrial visitations.

FINE-TUNING. In physics, the fact that fundamental forces have values within a relatively narrow range that would permit biology. In cosmology, the fact that the expanding universe has particular physical properties that allow biology to develop.

FREE RADICAL. An atom or molecule with an unpaired electron in its outer shell. Biologically relevant free radicals are highly reactive and are implicated in theories of cell aging by oxidative damage.

FUTUROLOGY. The study of, and prognostication about, the future. The longer the timespan, the less certain the prediction.

GAIA. The idea that the biosphere is a self-regulating system, where changes in the environment adjust so as to sustain life. A more controversial form of the idea posits Earth as a living organism.

GENOME. The hereditary information of an organism contained in its DNA or, for some viruses, RNA.

GENOTYPE. An organism's full hereditary information or its genetic makeup, even if not fully expressed.

GLOBAL WARMING. The upward trend in temperature over the past century, correlating strongly with human activity and the growth in the concentration of the greenhouse gas carbon dioxide.

GRAVITY TRACTOR. The use of spacecraft to fly alongside a threatening asteroid or comet and "steer" it by gravity onto a safe path.

GRAVITY WAVES. Ripples in space-time caused by any massive or compact object changing its configuration. Can also be called gravitational waves.

GREAT DYING. Term for the dramatic loss of biodiversity of the present era, a sixth mass extinction caused by human activity.

GREAT SILENCE. The fact that no intelligent civilizations have visited or communicated, despite the likelihood that they are present in the Milky Way galaxy.

GREEN GOO. The idea that some form of current or future microbe might be able to outcompete all other life-forms and take over the biosphere.

HABITABLE PLANET. The criteria by which a planet other than Earth may be able to host life, bounded by the limits of microbial life on Earth.

HABITABLE ZONE. Traditionally, the range of distances from a star where water can remain liquid on the surface of a planet or a moon. However, if planets and moons have internal sources of heat, the zone can expand markedly.

HADEAN. The period from Earth's formation until 3.8 billion years ago, during which the crust and oceans formed and life probably first started under challenging physical conditions.

HAWKING RADIATION. According to a theory by Stephen Hawking, black holes have a temperature and emit energy and matter, evaporating at a rate that increases as the mass of the black hole decreases.

HAYFLICK LIMIT. The natural limit to cell life, where a cell in a normal culture can only divide 50 times.

HORIZON. In cosmology, the fact that there is a limit to vision imposed by the finite age of the universe, so we are limited to seeing the places from which light can have reached us in 13.7 billion years and the total expanse of space is likely to be much larger.

HYPERNOVA. The most extreme cosmic explosion known, releasing a torrent of gamma rays and lethal radiation in twin jets. Lethal to the Earth if pointed at us within 1000 light-years.

IMMORTALITY. Living in tangible or intangible form for an infinite length of time. Some simple species are biologically immortal.

INFLATION. The first tiny fraction of a second saw an exponential gain in the size of the universe, rendering our visible region of space-time smooth and flat.

INTERSTELLAR TRAVEL. Travel between the stars. It requires technology of propulsion far beyond current capability to be able to go even to the nearby stars in a human lifetime.

LIFE EXPECTANCY. Usually quoted at birth, so the average can be skewed downward when there is high infant mortality.

MASS EXTINCTION. A sharp drop in the number of species in a short period of time, affecting all taxonomic groups. There have been five in the past half billion years of well-sampled fossils.

METABOLIC RATE. Measured per unit time, either food consumption, energy released as heat, or amount of oxygen consumed.

MILANKOVIĆ CYCLES. The sum of all the periodic effects of variations in the Earth's orbit on its climate.

MILKOMEDA. The hypthesized eventual state of the Milky Way, after the Milky Way merges with Andromeda in a few billion years.

MILKY WAY. The system of stars of which the Solar System is a part, containing about 400 billion stars.

MORTALITY RATE. The number of deaths in a population scaled to the size of the population, per unit time. Usually quoted as deaths per 1000 individuals per year.

MULTIVERSE. Hypothetical parallel universes that result from the initial quantum space-time that spawned our universe. They would all have different properties and physical states.

NEUTRON STAR. End state of a massive star where the remnant is of nuclear density and composed of pure neutrons.

NUCLEAR WINTER. The effect of a global or regional nuclear war, where deposition of particles into the upper atmosphere could cool the planet and damage the ozone layer.

OORT CLOUD. The reservoir of the comets that occasionally visit the inner parts of the Solar System. Not directly observed, it extends 50,000 times the Earth-Sun distance and contains a few trillion comets.

PANDEMIC. An epidemic that spreads uncontrollably through a large region, due to a communicable disease.

PANSPERMIA. The origin of life on Earth by transmittal from another cosmic source, like a neighboring planet or star system.

PARTICLE DECAY. Normal matter is predicted to decay in most theories on very long timescales, 10^{35} years or longer, yielding electrons, positions, and photons.

PATHOGEN. A biological agent that causes disease or illness to its host, can be bacterial or viral in nature.

PHENOTYPE. An organism's actual observed properties, including its behavior, morphology, and development.

PHYLOGENY. The history of organisms as they change with time. It is usually represented as a "tree of life" based on gradual changes in the genetic material.

POPULATION BOTTLENECK. Situation where disease, maladaptation, or environmental change leads a species to the brink of extinction. Genetic drift increases within the smaller population.

PROKARYOTE. Cell without a nucleus, the earliest and simplest form of microbial life on Earth.

QUASAR. Energetic phenomena near a supermassive black hole at the center of a galaxy that far outshine the starlight from the rest of the galaxy.

RARE EARTH. The idea, popularized by Peter Ward and Don Brownlee, that the conditions needed to develop complex life or intelligence are rare, and there will be few other planets like Earth.

RNA WORLD. Hypothetical and transition stage in the evolution of life preceding DNA and cells, where RNA coded genetic information and catalyzed its own replication.

SELF-REPLICATING SPACE PROBES. Also called bots, or von Neumann machines, they can travel to nearby stars, mine materials to make replicas of themselves, and explore the Galaxy.

SENESCENCE. All of the biological processes of an organism that lead to aging. The complexity of these processes, many of which are poorly understood, has led to many theories of aging.

SIMULATION HYPOTHESIS. Motivated by the likelihood that there are other intelligent civilizations much more advanced than ours, the proposition that we are living in a computer simulation created by a superior intelligence.

SINGULARITY. Hypothetical time in a few decades when exponential progress in nanotechnology, genetic engineering, and computing leads to a post-biological race of humans.

SNOWBALL EARTH. A time 700 million years ago, and probably also 2.2 billion years ago, when runaway climate change left Earth in a cold state where ice covered much of the planet.

SPACE COLONIZATION. The idea of living beyond Earth, either in orbit or on the surface of another planet or moon.

SPECIES. A basic biological definition, but surprisingly hard to define. A group of individuals able to interbreed and produce fertile offspring. It can also be defined using DNA or morphology.

STELLAR CORPSE. Any star after the final fusion reactions have ceased. Includes white dwarfs, neutron stars, and black holes, all of which either are compact end states or faded stars.

STERILIZING IMPACT. Space debris impacts early in Earth's history that could have eradicated all life on land and in the oceans. It's not known when the last such impact might have occurred.

STROMATOLITES. The fossilized residue of microbial colonies, extending to 3.5 billion years ago, with descendants thriving today.

SUPERCENTENARIAN. Someone who has lived to the age of 110 or more. Only 1 in 1000 centenarians lives to be this old.

SUPERFLARES. Occasional events where a Sun-like star has episodes of activity millions of times stronger than a normal solar flare.

SUPERMASSIVE BLACK HOLE. Formed at the center of galaxies, with a mass that scales with the stellar mass near the center. When the black hole is inactive it only reveals itself by its gravity on stars near the center. The power source of a quasar.

SUPERNOVA. The violent death of a massive star. Only a danger to Earth if it occurs within 25 light-years.

SUPERSTRING THEORY. Fundamental theory of nature that posits 10-dimensinal space-time, with particles as manifestations of tiny 1-dimensional strings.

SUSPENDED ANIMATION. The idea of bringing a metabolism down to a state close to death, as a way of surviving a long space journey.

TELOMERE. Repetitive DNA at the end of chromosomes, protecting them from cancer and destruction. Shortening of telomeres in cell division causes aging at a cellular level.

TERRAFORMING. Usually in reference to Mars, the idea of rendering a planet or moon fit for human habitation by altering its climate.

THANATOLOGY. The interdisciplinary study of death, usually in a cultural or scientific context.

TRANSHUMANISM. An international intellectual and cultural movement supporting the use of technology to improve human capabilities.

WHITE DWARF. The end state of the Sun and all low-mass stars. With no more nuclear fuels, white dwarfs are slowly cooling embers.

ZIRCON. A dense and durable crystal, examples of which date back 4.4 billion years. These ancient samples of Earth's crust indicate the planet had liquid water then and could have hosted life.

Notes

Chapter 1. Endings Are Personal

1   Western cultures have created masks of the recently deceased since the Egyptians routinely included a sculpted mask in the sarcophagus of each Pharaoh. Wealthy Romans used wax to preserve a likeness of family members, which was then used to create stone sculptures. Roman masks were artistically modified when they were rendered in stone to be more serene and heroic, but in the Middle Ages the tradition began of making masks based on wax or plaster molds and so true replicas of the face after death. Masks were made of famous people, not just royalty and rulers; examples include Dante, Voltaire, Chopin, and Keats.

2   Video games like "Grand Theft Auto" get the most publicity for their gratuitous violence, but ordinary TV has also become permeated with death and mayhem. The National Cable Television Association's "National Television Violence Study" found violence in 60 percent of all TV programs, two-thirds of which have bad characters that go unpunished, and only one-fourth of which show negative long-term consequences. The link between TV violence and aggressive and violent behavior in young adults has been demonstrated many times, most convincingly in a longitudinal study from the University of Michigan in 2003. Desensitization seems to be an unhealthy way to deal with the reality of death.

3 Data come from the "U.S. Religious Landscape Survey" of 35,000 adults in 2008 by the Pew Forum on Religion and Public Life. A survey by the American Association of Retired People in 2008 found that the belief is stronger among women than men and, perhaps unsurprisingly, belief grows with age after 50.

4 From an interview in the *Atlanta Journal-Constitution*, October 9, 2008.

5 From a letter by Boswell to William Johnson Temple in July 1858, published in C. B. Tinker, *Letters of James Boswell, Vol.1* (Clarendon Press, 1924).

6 From Ann Druyan's epilogue in Carl Sagan, *Billions and Billions* (Ballantine, 1997).

7 Clonal colonies complicate the debate over age. In long-lived plant and fungal species, no individual part of the colony is alive, in the sense of having a metabolism, for more than a small fraction of the age of the entire clone. Some clonal colonies are joined by their root systems, such as 40 hectares of Quaking Aspen in the Wasatch Mountains of Utah, and others are the vegetative parts of an asexually reproducing fungus, such as a 971-hectare, 2200-year-old mycelium in eastern Oregon. Ages for clonal colonies are estimates.

8 Data are culled from a nature bulletin of the Forest Preserve District of Cook County, which collects reliable records from zoos and aquariums all over the world.

9 The U.S. Census Bureau issues a report each decade on the profile of American centenarians. In that study and others that have looked for common features among those who live to be 100, the answers are fairly simple: don't smoke, get lots of sleep and exercise, eat a balanced diet, and stay mentally engaged. Other things to avoid are being overweight, getting divorced, and excessive stress; centenarians seem to have the skill of not internalizing stress. Good genetics helps, but you have no control over that. Plus advice from fictional centenarian Huck Finn, "today is what counts, fine clothes are itchy, and money is a burden."

10 The Pew Forum on Religion and Public Life conducts regular surveys on the U.S. religious landscape. Americans are very religious but nondogmatic; most agree with the statement that many religions, and not just their own, can lead to eternal life (atheists and agnostics are on their own . . .). The most recent survey in 2008 found that 6 in 10 Buddhists believe in Nirvana and the same proportion of Hindus believes in reincarnation. For all the major religious traditions, belief in Heaven is stronger than belief in Hell.

11 Descartes' skepticism caused him to worry that all of his experiences might be the result of a powerful outside force: a "malicious demon." This radical worry seems inescapable; how could you prove that you are not in the nightmarish situation Descartes describes? Modern philosophers have extended this idea, by asking how you could prove that you are not just a "brain in a vat," and there is an echo of Descartes in the premise for the movie *The Matrix*. There's no neat way out of this problem but Descartes argued that one cannot doubt the existence of oneself. All thinking presupposes a thinker; even in the act of doubting, there must be a self doing the doubting. Hence his famous line: "I think, therefore I am."

12 The twentieth century saw the rise of biological explanations of life and mechanistic explanations for brain function. Gilbert Ryle famously derided dualism as "Descartes' myth" and "the dogma of the Ghost in the Machine." He said dualism stemmed from thinking what philosophers call a category mistake. Suppose you visit a university and get a full campus tour, seeing the dorms, the classrooms, the library, and so on. At the end of the tour, you ask, "Where is the university?" But the university is not a separate entity apart from all its constituents; it stands for the entire collection. Similarly, Ryle argued, the mind should not be thought of as some separate entity from the body, or brain. While ascendant, materialism is not triumphant; scientists have not, for example, successfully explained consciousness in terms of physical brain function.

13 Published in *Lancet*, December 15, 2001. See also Gary Habermas, "Near Death Experiences and the Evidence: A Review Essay," *Christian Scholar's Review* 26 (1996), p. 78.

14 Karl Jansen, "The Ketamine Model for the Near Death Experience: A Central Role for the NMDA Receptor," http://leda.lycaeum.org (retrieved in December 2008).

Chapter 2. All Good Things Must Pass

1 Although it is one of the oldest documents in human history, few current doctors swear to the traditional Hippocratic Oath at the end of their training. Many medical schools use a modern version written in 1964 by Louis Lasagna, Dean of Tufts Medical School. The newer oath omits prohibitions against euthanasia and abortion, but it also conveniently fails to prohibit sexual contact between doctor and patient.

2 Data from U.S. Department of Commerce, Bureau of the Census, *Historical Statistics of the U.S.*

3 Data from U.S. Public Health Service, *Vital Statistics of the United States*, vols. I and II, part of the National Vital Statistics System.

4 Data from the World Heath Organization, Geneva, Switzerland, *World Heath Statistics Annual.*

5 Data from the World Health Organization, *International Accident Facts*, 3rd ed., and *World Report on Violence and Health.*

6 *Condé Nast Traveller* magazine carried an article in February 2003 that polled readers on their greatest travel fears and compared them to the statistical risks.

7 The National Safety Council gathers and publishes statistics on modes of death and injury from the mundane to the highly obscure, based on data from the National Center for Health Statistics and the U.S. Census Bureau.

8 Death rates parsed by profession have always been difficult to interpret, and the statistics on suicide are even more controversial because there is no national database. The National Institute of Occupational Safety and Health reported in 1995 that members of the medical profession have higher suicide rates, as do black, male guards and policemen, and white female artists. Profession is not strongly predictive; psychologists have long documented that the top predictors of suicide are mental disorder, substance abuse, loss of social support, and easy access to a firearm.

9 In 2009, the Nobel Prize in physiology was awarded to Elizabeth Blackburn, Carol Greider, and Jack Szostak for their work in discovering how telomeres protect chromosomes.

10 The fundamental definition of entropy is statistical; it describes the number of possible microscopic states of a system. German Physicist Ludwig Boltzmann had the equation inscribed on his tombstone. The connection between microscopic disorder and heat is clear in the example of ice melting into water, where the ordered array of molecules in ice is converted into disordered and freely flowing molecules of water. Life uses energy, and negative entropy is a measure of the usefulness of energy. Biological processes take in negative entropy in the form of energy-storing molecules and releasing a larger amount of entropy as heat into the environment. A moderately gentle introduction to thermodynamics and life can be found in J. Miguel Rubi, "Does Nature Break the Second Law of Thermodynamics?" *Scientific American*, October 2008.

11 The science of cryonics is rapidly evolving, which means the hypothetical viability of the earliest frozen corpses is in even greater doubt. Cryonics embeds a "last-in, first-out" strategy, where future methods of resuscitation will work the best on the most recently interred patients. University of Helsinki researcher Anatoli Bodgan has experimented with "glassy water" that can be slowly supercooled to an amorphous state without crystallization and cell damage. Unsurprisingly, in our litigious age, cryonics has spawned a new branch of law. *The Wall Street Journal* reported on January 21, 2006, about "frozen trusts" that allow cryonic patients to have their estates managed and investment decisions made while they are on ice. Their assets don't pass to their descendants; patients gain the benefit of thousands of years of interest when they're resuscitated.

12 The definition of death is of great interest to the world's religions, who would like the medical profession to adopt procedures in accord with their teachings. For example, the Pontifical Academy of Sciences held a conference in 2006 with doctors and neurologists from around the world, to explore the definition of death. The resulting recommendations and discussion appeared in "Signs of Death," *Scripta Varia*, Vol. 110 (Pontifical Academy of Sciences, 2007).

13 As we saw in the last chapter, the increase in the number of successful resuscitations has spawned research and a growing literature on near-death experiences. In 2008, researchers at the University of Southampton launched the "Awareness During Resuscitations" study involving patients at hospitals in Europe and the United States. About 20 percent of cardiac arrest patients have a near-death experience where they recall details of the surgery, and sometimes it can be well-documented. If there is a "third state" of being between living and dying, it's probably good news for cryonics patients.

Chapter 3. The Future of Humanity

1 Carolus Linnaeus is not as famous as he should be, given his pivotal role in botany and zoology. The philosopher Jean-Jacques Rousseau sent a message saying, "Tell him I know no greater man on Earth." Linnaeus developed the binomial classification of species that is still used, and even tried to extend it to minerals. He was a great teacher and inspired many students, although more than a few died collecting specimens for their mentor.

2 Microbes present a real headache for classification, because the traditional species concept fails when there is cloning or asexual reproduction, and when morphology is no guide. Microbiologists have embraced a definition based on genetic overlap, such that species must share at least 70 percent DNA overlap or 97 percent gene overlap for 16S ribosomal RNA (rRNA). But W. Ford Doolittle has pointed out in *Microbiology Today* (November 2006, p.148) that lateral gene transfer is prevalent among prokaryotes (life-forms composed of a simple cell with no nuclei) and it complicates any genetic definition of species separateness.

3 The situation where a physical barrier prevents breeding and leads to genetic divergence into separate species is called allopatric speciation. A variant of this occurs when the geographically isolated population is much smaller than the main population; Ernst Mayr identified this "founder effect" in a small population; it can lead to genetic drift and what is called peripatric speciation. A third example is two populations that are physically adjacent and only have a small zone of contact in a continuous habitat; variations in mating frequency lead to what is called parapatric speciation. The most controversial mode of speciation occurs when genetic divergence occurs within the same geographical area, called sympatric speciation. "Resident" and "transient" orca forms in the Pacific Northwest provide an example.

4 The barrier has already been breached in ways that some might find unsettling. In April 2008 British newspapers reported that a team led by Lyle Armstrong created the first hybrid human-animal embryos. They inserted human DNA from a skin cell into a hollowed out cow egg, and used an electric shock to induce it to grow. The embryo, 99.9 percent human and 0.1 percent animal, grew for three days until it had 32 cells. The goal is to harvest such hybrids for stem cells.

5 Philosophers view the flexibility of the definition of species sternly. They question its ontological status and wonder whether species are kinds of creatures or individuals. They see no sign that biologists have decided between species monism and pluralism—whether there is one biological principle that unites all members of a species or not. Finally, they are dubious whether the term "species" is a real category in nature.

6 In June 2008, the Spanish Parliament passed a resolution to grant primates the rights to life and liberty. This followed Switzerland, in 1992, and Germany, in 2002, recognizing animals as beings not things. The movement to grant great apes "personhood" has gained momentum in the developed world, and it's supported by prominent people such as Jane Goodall, Richard Dawkins, Princeton philosopher Peter Singer, and Harvard law professor Steven Wise. It does, however, raise some complex legal and ethical issues. If apes have rights, do they also have responsibilities? What about situations in the wild where chimps commit infanticide or castrate their neighbors or torture an antelope they have no intention of eating? We define the rules but must admit that the issue of personhood is not black and white. We will never be able to experience the interior mental landscape of another creature.

7 The corollary of the fact that morphologically diverse species share much genetic material is the fact that features that mark one human "tribe" out from another stem from the action of a handful of genes. Most of the distinctive features of race—skin color, hair color and texture, facial features—correspond to genetic diversity smaller than would be found among members of any one race.

8 The suggestion of hybridization between chimps and humans is controversial because it depends on the reliability of time using DNA sequences, which can be less reliable than radioactive dating of a fossil skull. The principal finding of the genetic analysis from a team led by David Reich from the Broad Institute in Cambridge, Massachusetts, is the X chromosomes of humans and chimps diverging about 1.2 million years later than the other chrómosomes (females have two X chromosomes, males one X and one Y).

9 Ajit Varki and a team at the University of California at San Diego found a gene facilitating brain growth in people, but not apes, which entered the human lineage via a mutation about 2.7 million years ago.

10 Research done by Patrick Evans and collaborators at Howard Hughes Medical Institute, published in *Science* 309 (2006), p. 1717. The same team had earlier identified 45 advantageous amino acid changes in *microcephalin* over 30 million years of evolution from early simians to modern humans.

11 It is not trivial to measure the background extinction rate because short-lived species may be imperfectly sampled from the incomplete fossil record.

12 Data from the U.S. Environmental Protection Agency, Municipal Solid Waste "Fact Sheet," 2007, at http://www.epa.gov/waste/nonhaz/municipal/msw07-fs.pdf.

13 Data from the Center for Defense Information, Washington, D.C., part of the World Security Institute (2006).

14 Interview at the State of the World Forum in New York, September 7, 2000, published in *Salon*, www.salon.com.

15 Interview published in the MIT magazine *Technology Review*, March 2006. Popov's credibility is an issue; as a defector he would have motivations to offer "important" revelations to his new adopted country. However, many elements of his stories are corroborated by other scientists from Russia, and the scientific aspects are considered plausible by Western scientists. Popov was extensively interrogated and debriefed when he defected, and another revealing interview is on the U.S. Homeland Security Institute Web site, www.homelandsecurity.org.

16 Commission on the Prevention of WMD Proliferation and Terrorism report, "World at Risk," released December 3, 2008, www.preventwmd.gov.

Chapter 4.Beyond Natural Selection

1 Richard Fortey, *Earth: An Intimate History* (New York: Vintage, 2005).

2 Jason Bond, a biology professor at East Carolina University, had previously named a spider after left-wing rocker Neil Young. Colbert complained on national TV and demanded tribute, and in June 2008 he got his way, with *Aptostichus stephencolberti*.

3 From a report released by the National Geographic Society in April 2008, based on research published in the *American Journal of Human Genetics* 78 (2006), p. 487.

4 Joshua Lederberg, "The Microbial World Wide Web," *Science* 288 (2000), p. 291.

5 A first version of this tool was rolled out in 2008 and it's a brilliant example of the powerful data that Google has at its fingertips. Google Flu follows conventional flu outbreaks across the U.S. by tracking keyword searches on flu symptoms or remedies. The map is updated daily; by contrast the Centers for Disease Control, which relies on visits to doctors' offices and similar data, has a time lag of two weeks.

6 From a lecture at University College, London, October 7, 2008, quoted in the *London Times* October 8, 2008.

7 Ray Kurzweil was not the first to speculate about the singularity. Vernor Vinge, a mathematician at San Diego State University, presented the idea at the VISION-21 Symposium in March 1993, and an abridged version of it was published in the Winter 1993 issue of the *Whole Earth Review*.

8 Bill Joy, "Why the Future Doesn't Need Us," *Wired*, April 2000. It sparked a rebuttal from Kurzweil in his 2005 book *The Singularity Is Near*, and a critique by John Seely Brown and Paul Duguid, *AAAS Science and Technology Policy Yearbook* (Washington, DC: American Association for the Advancement of Science, 2001), "A Response to Bill Joy and the Doom-and-Gloom Technofuturists."

9 In fact, the product of a series of factors is as uncertain as the most uncertain factor, so precision in some of the astronomical numbers early in the Drake equation is no help if the later, sociological factors are very speculative. One sample of life is not enough to apply induction to the expectations for a large universe.

10 If there is a long tail of civilizations of great longevity, it will skew the average longevity to a higher value. For example, a single civilization that lasts a million years counts as much as a thousand lasting a thousand years in the Drake equation. Reversing this logic, the airwaves might be silent except for a few essentially immortal civilizations.

Chapter 5. The Web of Life

1 If they were particularly unlucky, one of the probes might land in the Atacama Desert of northern Chile. A place of salt pans, volcanoes, and the largest copper mine on the planet, it's the place astrobiologists go when they want to find a Mars-like habitat on Earth. In 2003, a group led by Rafael Navarro-Gonzalez tested the soil in a particularly arid region using equipment like that on the Viking Mars lander and were unable to detect DNA. As opposed to Mars, where we will have to look hard to find life, on Earth we have to look hard *not* to find life.

2 It's hard to imagine how a tiny chip of natural crystal can speak to the condition of an entire planet just after it formed, but zircons from the Jack Hills formation of Western Australia have concentrations of oxygen isotopes that suggest liquid water on Earth 4.4 billion years ago. If the planet had continents, weathering, and relatively cool temperatures that long ago, it might also have hosted biology.

3 Pasteur was a master using the experimental method to cut through ambiguity and test hypotheses with elegance and efficiency. Spontaneous generation was finally laid to rest in 1859 when the French Academy held a competition to test the theory. Pasteur filled flasks with beef broth and

boiled them, leaving some flasks open to the air and sealing others. The open flasks became contaminated with microorganisms while the sealed flasks did not. Then he put boiled broth in flasks with long S-shaped necks; air could enter them but microorganisms were trapped in the bend in the neck. The flasks were uncontaminated months later.

4 DNA evidence suggests the earliest common microbial ancestor was a mesophile, a microbe happy near but under the boiling point of water. However, the crust cooled quickly and some environments were very cold. Researchers have explored an origin of life under near-freezing conditions. In February 2008, *Discover* magazine reported on a time 10 years earlier when Stanley Miller had shown that seven different amino acids and 11 different nucleotide bases had formed at −78°C (−108°F), the temperature of Europa. The same article also reported that Hauke Trinks had formed RNA molecules 400 bases long under freezing conditions—at low temperatures the ice crystals form pure lattices, squeezing out impurities and speeding templating reactions.

5 Antonio Lazcano and Stanley Miller, "How Long Did It Take Life to Begin and Evolve to Cyanobacteria?" *Journal of Molecular Evolution*, 39 (1994), p. 546.

6 Prokaryotes are the simplest form of cell. They don't have a nucleus and they formed first in the evolution of life on Earth. Bacteria and Archaea are the two largest categories of prokaryotic cell. Eukaryotes are more complex cells and include plants, animals, fungi, and protists. Their first appearance is uncertain, but was probably 2 billion years ago and may have been as long as 2.7 billion years ago.

7 Conversely, the presence of life in another of Chris MacKay's favorite haunts—the high and dry valleys of Antarctica—spurs optimism that life might be found on Mars. In 2003, a team led by William Mahaney found fungi and bacteria several inches under the hard and parched surface of several Antarctic valleys, Astrobiology Magazine online, July 11, 2002.

8 From an interview with Diana Northup on the PBS Web site for NOVA's "The Mysterious Life of Caves," retrieved December 2008 from www.pbs .org/wgbh/nova/caves.html.

9 Quake is a brilliant example of a scientist who saw his best chance to make a mark at the junction of the traditional fields of physics, biology, and engineering. Call him a nano-plumber if you will. He has used semiconductor lithography techniques to put thousands of valves, channels, and wells on a

single chip—the complexity of a major chemical factory reduced to the size of a postage stamp.

10 From Carol Cleland and Shelley Copley, "The Possibility of Alternative Microbial Life on Earth," *International Journal of Astrobiology* 4 (2005), p. 165.

11 Research by Robert Hazen and collaborators, "Mineral Evolution," *American Mineralogist* 93 (2008), p. 1693.

12 From an interview with James Lovelock in the *London Guardian*, March 1, 2008.

13 In the late 1980s Joe Kirschvink, a geophysicist at Caltech, found evidence for glaciation at low latitudes. He added a powerful line of evidence to the argument, based on geomagnetic data. The latitude at which rocks form can be estimated from the inclination of their natural magnetism, and the magnetic orientation confirmed that the glaciated rocks had been equatorial.

14 The interplay of the biosphere with the atmosphere produces the strongest biomarkers when a planet is seen from afar with spectroscopy. However, a planet or moon without a substantial atmosphere could host life underground or underwater even if the surface were sterilized by ultraviolet radiation. Remote sensing would not reveal biosignatures; spectroscopy can only identify a subset of habitable or inhabited planets.

Chapter 6. Threats to the Biosphere

1 The random location of impacts is easier to understand than the random timing. Earth spins on its axis and orbits the Sun and meets objects arriving with a wide range of trajectories, ensuring that no place on the surface is more likely to get hit than any other. The timing is random in the sense that it is governed by statistics of rare events. If the mean time between devastating impacts is 100 million years, that just means half of the intervals between impacts are more than 100 million years and half are less, but the distribution of intervals peaks at zero, which means that most probably the time of the next impact is soon! On the other hand, there is a long tail of intervals so successive impacts may be separated by 200 or 300 million years. This is like the intervals between buses in a busy city, where officials can rightly claim that the bus is most likely to come "soon" while our experience is that it can take a long time.

2 Impact explanations for recent geological events are headline grabbing, but most geoscientists remain skeptical. For example, see Nicolas Pinter and

Scott Ishman, "Impacts, Mega-Tsunami, and Other Extraordinary Claims," *Geological Society of America Today* 18 (2008), p. 37.

3 Specifically, the distribution of meteor size and numbers (or impact rate) is an inverse power law, with logarithmically larger-number objects of smaller size, such that there are 10 times more objects of 10 times smaller size, and so on (see Figure 6.1). This relationship holds over 7 orders of magnitude from 0.1 kilogram to 10 million kilograms (22,000,000 pounds). A key attribute of power law distributions is the fact that there is not "typical" size for an object. Power laws are seen in a wide variety of natural phenomena in physics, geology, chemistry, and biology; their ubiquity is explored elegantly in a book by Danish physicist Per Bak called *How Nature Works: The Science of Self-Organized Criticality* (Copernicus, 1996).

4 David Pankenier, Zhentao Xu, and Yaotiao Jiang, eds., *Archaeoastronomy in East Asia: Historical Observational Records of Comets and Meteor Showers from China, Japan, and Korea* (Amherst, NY: Cambria Press, 2008).

5 The "Earth Impact Effects Program" was created by Robert Marcus, Jay Melosh, and Gareth Collins and is hosted on their Web site at the Lunar and Planetary Lab at the University of Arizona, http://www.lpl.edu/impact effects. A document on the site explains the assumptions, observations, and calculations that go into the program.

6 To keep both the risk and the cost in perspective, the full spectrum of protection strategies, from an array of telescopes to spot Earth-crossing objects to spacecraft to deflect or destroy them, would cost a few tens of billions of dollars to implement. This is a tiny fraction of the cost of mitigating global warming or getting the world economies out of the recession of 2008–2010.

7 N. Sleep, K. Zahnle, J. Kasting, and H. Morowitz, "Annihilation of Ecosystems by Large Asteroid Impacts on Early Earth," *Nature* 342 (1989), p. 139.

8 See, for example, Peter Ward, "Mass Extinction: The Microbes Strike Back," *New Scientist*, no. 2632, February 9, 2008.

9 J. Cisar and colleagues, "An Alternative Interpretation of Nanobacteria-Induced Biomineralization," *Proceedings of the National Academy of Sciences* 97 (2000), p. 11511.

Chapter 7. Living in a Solar System

1 Most of the exoplanets discovered in the decade after the first discovery in 1995 were Jupiter mass or larger; Jupiter is 318 times the mass of Earth. This

didn't reflect a paucity of smaller planets but the insensitivity of the Doppler detection technique to lower masses. Improvements in Doppler detection have taken the detection limit to the mass of Uranus or lower; Uranus is 15 times the mass of Earth. In the past five to six years, most of the lowest-mass exoplanets have been discovered by microlensing, where an unseen planet briefly amplifies the light of a passing background star, according to an effect that was predicted by Einstein's theory of general relativity. In April 2009, a planet called Gliese 581e was shown to have a mass of only 1.9 earths. The era of being able to detect clones of our planet has arrived.

2 Gott has popularized astronomer Brandon Carter's decades-old Doomsday argument. It used Copernican logic to say that our place in the history of all humans should not be special or unusual.

3 David Grinspoon in "The Rare Earth Debate, Part 3: Complex Life," www.space.com, July 22, 2002.

4 From a much-cited article by Cark Haub, "How Many People Have Ever Lived on Earth?" *Population Today* 30 (2002), p. 3. It rebuts an old claim that most people who have ever lived are alive today; the correct percentage is only 5 or 6.

5 Skeptics have countered that the assumption that we are not living in a special time in the history of all humans is a strong assumption and not as warranted as the assumption that our place in the universe isn't special. To say that a species will likely live twice its current age is simplistic and somewhat naïve because it should hold for tigers, polar bears, golden frogs, and many other species on the verge of extinction.

6 D. Valencia, D. Sasselov, and R. O'Connell, "Detailed Models of Super-Earths: How Well Can We Infer Bulk Properties?" *Astrophysical Journal* 666 (2007), p. 1413.

7 Even though the largest moons rival the smallest planets in size and mass, it's because their attendant planets are very large. Moons are always much smaller than the planets they orbit.

8 Robert Pappalardo in "Europa Mission: Lost in NASA Budget," www.space.com, February 7, 2006.

9 Adam Showman and Renu Malhotra, "The Galilean Satellites," *Science* 296 (1999), p. 77.

10 A case could be made for both numbers being closer to 100 percent. Life on Earth formed quickly and spread to almost every conceivable ecological niche while the planet was very inhospitable, consistent with life as a preva-

lent feature of habitable planets. Complexity and multicellularity may be contingent outcomes of evolution, but if they flow naturally via evolution over billions of years, then the number of planets with large-scale life forms may be in the billions, in just one galaxy among billions in the universe.

11 Hans Rickman and collaborators, "Injection of Oort Cloud Comets: The Fundamental Role of Stellar Perturbations," *Celestial Mechanics and Dynamical Astronomy* 102 (2008), p. 111.

12 Research from a team led by Gunther Korschinek, *Physical Review Letters* 93 (2004), p. 1170.

13 R. Casadio, S. Fabi, and B. Harms, "On the Possibility of Catastrophic Black Hole Growth in the Warped Brane-World Scenario at the LHC," arXiv:0901.2948, 2009. There's an even more implausible line of speculation, suitably relegated to this endnote, that scientists experimenting with matter could spawn "baby universes" from extra dimensions that would consume us and Earth. The theory and the mechanism for this are so uncertain that it is comparable to worrying that a hypothetical and exceptionally rare anti-Earth will collide with us and annihilate us in a flood of gamma rays.

14 Our ignorance over the causes and mechanisms of stellar cataclysm is one issue. Equally important is the problem of tying ancient celestial events to ancient terrestrial events. Extinctions of any kind have multiple possible explanations and it's very difficult to rule out geological or environmental changes because rock layering does not give an age precision much better than 10,000 years. That's not good enough to prove that the trigger was instant and astronomical rather than gradual and geological. When massive stars die the hot gas they eject eventually cools and dissipates and they leave behind black holes or neutron stars which are dark and difficult to detect. Stellar motions within the Milky Way take them far from their position in the sky when they exploded. Progress on this topic will probably be slow and controversial.

15 Adrian Melott and collaborators, *International Journal of Astrobiology* 3 (2004), p. 55.

16 Peter Tuthill and collaborators, *Astrophysical Journal* 675 (2008), p. 698.

Chapter 8. The Sun's Demise

1 Herschel's correlation between sunspot absence and high wheat prices has been mocked, debunked, and endlessly scrutinized, but it has refused to die. In 2003, Lev Pustilnik and Gregory Din published a paper with a statistical analysis that found the correlation that was 99.8 percent significant ("Influence of Solar Activity on Wheat Market in Medieval England," Proceedings of International Cosmic Ray Conference, p. 4131). Wheat is unusual because there are 800 years of unbroken price records from European markets. Intriguingly, wheat futures reached an all-time high in 2008 just as we approached a sunspot minimum.

2 For a long time after Milankovic's original speculation, data were not good enough to confirm the effects. Deep-sea cores became available in the 1970s, and a seminal paper by Hays, Imbrie, and Shackleton called "Variations in the Earth's Orbit: Pacemaker of the Ice Ages" established the field of astronomical effects on climate (*Science*, 1976, Vol. 194, p. 1121). There is no explanation for the transition 3 million years ago from erratic variations to the two major cyclic imprints. In principle, Milankovic theory can be used to predict future climate change, modulo human impacts. Berger and Loutre predict the current warm climate for another 50,000 years in "An Exceptionally Long Interglacial Ahead?" *Science* 297 (2002), p. 1287.

3 The future of white dwarfs is a matter of speculation because the universe is only 13.7 billion years old, so no white dwarfs have cooled for longer than that amount of time. In fact, the logic can be reversed so the cooling rate of white dwarfs is used to put a limit on the age of the universe and it's reassuringly consistent with estimates from expansion rate and the microwave background radiation. White dwarfs will not turn black for many trillions of years. Black dwarfs should not be confused with black holes, which are left over when massive stars die, and brown dwarfs, which are stars of such low mass (less than 8 percent of the Sun) that their temperature puts them below the threshold for nuclear fusion.

4 Stephen Hawking interviewed on BBC Radio 4, reported in the *London Daily Mail*, December 1, 2006.

5 C. Bennett and colleagues, "Teleporting an Unknown Quantum State via Dual Classical and Einstein-Podalsky-Rosen Channels," *Physical Review Letters* 70 (1993), p. 1985. It works like this. Allie (A) and Brad (B) share an entangled qubit of information (AB), which can exist in four states. C is the qubit that Allie wants to transmit to Brad. Allie applies a mathematical operation to the qubits AC and measures the result to get two classical bits, destroying the two qubits in the process. Due to entanglement, Brad's qubit now contains information about C, but the information is randomized and

Brad's qubit B could be in any one of four states, preventing him from getting any information about C. Allie sends Brad her two measured qubits that indicate which of the four states Brad holds. Brad applies a mathematical operation depending on which qubits Allie sends him, transforming his qubit into an identical copy of the qubit C. Voilà!

6 S. Olmschenk and colleagues, "Quantum Teleportation Between Distant Matter Qubits," *Science* 323 (2009), p. 486.

7 Altering Mars enough to sustain extremophiles is a huge undertaking with a hefty price tag. Altering it enough to support human habitation would be so expensive that it would usurp a substantial fraction of our global resources if we needed to do it within a century. Many people think the whole idea is wrong-headed. If we are facing problems with our own biosphere, it would be far cheaper to establish sealed habitats and domed cities on the surface or underground on Earth, rather than on Mars. Shipping more than a hundred or so colonists to Mars would also be prohibitively expensive. Without unforeseen technologies to assist us, our best bet for the immediate future is to keep Earth habitable or find a new safe haven here.

8 Technology mavens worry about how it can be done, but rarely ask if it should be done. The ethical considerations of irrevocably altering a planet for our purposes deserve consideration too. Even among scientists, there's a spectrum of views. Chris McKay thinks it's our right and obligation to spread biology to lifeless places, while Woodruff Sullivan, an astrobiologist at the University of Washington, thinks alien worlds should be preserved in their natural states, like national parks.

9 D. Korycansky, G. Laughlin, and F. Adams, "Astronomical Engineering: A Strategy for Modifying Planetary Orbits," *Astrophysics and Space Science* 275 (2001), p. 349, and report in *The New York Times*, June 17, 2001.

10 Donna Haraway, "A Cyborg Manifesto: Science, Technology, and Socialist-Feminism in the Late Twentieth Century," in *Simians, Cyborgs, and Women: The Reinvention of Nature* (New York: Routledge, 1991), p. 149.

11 A. Sandberg and N. Bostrom, "Whole Brain Emulation: A Roadmap," *Technical Report* #2008-3, Future of Humanity Institute, Oxford University.

Chapter 9. Our Galactic Habitat

1 Precisely because the sky is mostly unchanging, temporal variations have been very significant for human cultures. The planets move among the pat-

tern of fixed stars so they have been the subject of myth and divination for thousands of years. The rare occasions a star visibly brightens or becomes a supernova make deep impressions on people who have no scientific explanation for the phenomena. The great distances to stars mean their motions are not visible over a human lifetime, so constellations are cultural "artifacts" that get handed down from generation to generation.

2 Despite Aristotle's dismissal of the idea, his work informs us that Anaxagoras and Democritus both thought that the Milky Way might consist of distant stars. The history of early speculation about the Milky Way highlights the important role of Arab thinkers. In the early tenth century, the Persian astronomer Abu Rahyan Al-Buruni proposed that the Milky Way was made of individual points of light where refraction in the Earth's atmosphere blurred each star's light. Three hundred years later, Ibn Qayyim Al-Jawziyya thought that the Milky Way was crammed with sources of light, each of which was larger than a planet.

3 The Milky Way provided the first indication that the universe was dominated by a dark form of matter. Looking outward through the disk, astronomers can map the velocity of stars as they orbit the disk; gravity theory predicts that the speeds should decline as the periphery of the Galaxy is approached, much as the speed of outer planets declines with distance from the Sun. Instead, the orbital speeds remained constant out to the visible edge of the Galaxy, indicating gravity from unseen matter driving the motions. The same effect has now been seen in hundreds of galaxies of all types—dark matter is a ubiquitous feature of the universe. In the Milky Way, observations are sensitive enough to rule out contributions from dust, rocks, failed stars (or brown dwarfs), and black holes. This leaves massive but weakly interacting subatomic particles as the most plausible explanation.

4 Although it sounds exotic, evidence for the massive black hole in the center of the Milky Way is as good as evidence for more prosaic black holes that form when massive stars die. Orbital motions of stars near the center give a robust mass estimate, and the concentration of matter is too high to be caused by a star cluster. Current observations using the Keck Telescope in Hawaii aim at demonstrating for the first time that black holes are surrounded by an event horizon, marking the limit of vision and information.

5 The Millennium Simulation graced the cover of *Nature*; see Volker Springel, Simon White, Carlos Frenk, and collaborators, "Simulations of

the Formation, Evolution and Clustering of Galaxies and Quasars," 435 (2005), p. 629.

6 The technical issues and limitations of the simulations are still significant. Structure formation is guided by the dominant dark matter, but including normal matter is essential because that's what gives visible stars and galaxies. Yet the complex astrophysics of star formation can't yet be included in a simulation so the properties of galaxies have to be fed in "by hand." Other simulations attempt to include gas and be more realistic about how galaxies actually form. Although the paradigm for structure formation is "top down," meaning that small objects form first and larger objects later, the simulations have to explain an observed phenomenon called "down-sizing," where large galaxies go through their most active star formation early and small galaxies are only forming stars recently. Given the challenges, simulations agree fairly well with observations and support the cosmological model where dark matter and dark energy govern the cosmic expansion rate.

7 One remaining puzzle, apart from the nature of dark matter itself, is the reason the Milky Way looks so "lumpy" in dark matter and why few of the small lumps of dark matter in the universe contain any visible stars. Astronomers speculate that gas falling into the many smaller pockets of dark matter stays too hot to collapse into stars, leaving those regions dark.

8 Understanding nuclear fusion starts with a counterintuitive fact: the sum of the masses of the two protons and two neutrons that make up a helium nucleus is less than the mass of the four particles that went into it. How can the whole be less than the sum of its parts? A helium nucleus is stable, which means it doesn't spontaneously split into its constituent parts. In fact it takes energy to break it apart against the forces that keep it together. Energy doesn't spontaneously appear or disappear so that energy before is the same as the energy after. That means the energy of the helium nucleus plus the energy needed to split it up equals the energy of its separate parts. By simple arithmetic the energy of the helium nucleus equals the energy of its parts minus the energy needed to split it up. That negative energy is called the binding energy, and because Einstein showed that mass is equivalent to energy, an energy deficit is equivalent to a (tiny) mass deficit. Hence the whole is less than the sum of its parts.

9 T. J. Cox and A. Loeb, "The Collision Between the Milky Way and Andromeda," *Monthly Notices of the Royal Astronomical Society* 386 (2007), p. 461.

10 A. Ghez and collaborators, "Stellar Orbits Around the Galactic Center

Black Hole," *The Astrophysical Journal* 620 (2005), p. 744; and R. Genzel and collaborators, "The Stellar Cusp Around the Supermassive Black Hole in the Galactic Center," *The Astrophysical Journal* 594 (2003), p. 812.

11 P. Hopkins and collaborators, "A Cosmological Framework for the Co-Evolution of Quasars, Supermassive Black Holes, and Elliptical Galaxies. I. Galaxy Mergers and Quasar Activity," *The Astrophysical Journal Supplement* 175 (2008), p. 356.

12 Astronomers just see a snapshot of activity in galaxies at different epochs; they don't get to follow the evolution of any particular galaxy. Surveys show that the peak of the quasar era was 10 billion years ago when they were hundreds of times more abundant than they are now. The centers of many large galaxies must currently be starved of gas, as the Milky Way is now. In the nearby universe, only 1 in 1000 bright galaxies hosts a quasar. Since there is evidence that all bright galaxies harbor massive black holes, they must only spend a thousandth of their time in the "on" state, or roughly 10 million years at a time.

Chapter 10. Aging of the Milky Way

1 The reason is that nuclear fusion rates are very sensitive to temperature. More massive stars have more gravity, which creates higher core temperatures, driving fast and efficient fusion. The rate of consumption of fuel far outweighs the extra fuel available in a more massive star, which is why massive stars have short lives compared to low-mass stars. In any population of stars, the massive and hot ones die the quickest, so the population gets dimmer and redder as it ages.

2 These proportions are established by the mass distribution of stars as they form, which is thought to be a standard power law relation first pointed out by Princeton astrophysicist Ed Salpeter. The paucity of high-mass stars at the time of formation converts into a paucity of neutron stars and black holes after the energy sources have been exhausted.

3 Modern cosmology has as a foundation the assumption that the laws of nature don't vary from one place to another or one time to another. Edwin Hubble gave a nod to the assumption of the "uniformity of nature" when he said that his determination of the distance to Andromeda depended on the fact the Cepheid stars behave the same in our galaxy as in any other galaxy. Cosmologists accept that physical conditions in the universe change with

time and place—temperature, density, pressure, chemical composition—but they assume that the underlying physical constructs don't change. If they do, understanding the universe becomes a lot more difficult. On the other hand, the universe need not be comprehensible just for our benefit.

4 For the entire 400-year history of modern astronomy we have viewed the universe in electromagnetic radiation, using optical telescopes and, in the past 50 years, telescopes at longer and shorter wavelengths. But what would the universe look like if we had gravity "eyes"? We'll find out as the advanced version of the Laser Interferometer Gravitational Wave Observatory begins looking for gravity waves in 2014. This exquisitely precise instrument will be able to detect binary neutron stars merging out to a distance of 1 billion light-years and binary black holes merging out to a distance of 5 billion light-years. Thousands of galaxies' worth of stars are within the detection zone so even if the events are rare they should be detected weekly or even daily.

5 Businessmen might be excused for keeping their eye on the bottom line and so underplaying the future, but the predictions of science fiction writers are not much more reliable. For every example like Arthur C. Clarke predicting telecommunications satellites there are many hard science fiction writers who predicted space-faring civilizations that we have yet to see or emulate. Perhaps it's more useful to remind ourselves of Clarke's three laws of prediction: (1) When a distinguished but elderly scientist states that something is possible, he's almost certainly right; but when he states that something is impossible, he's almost certainly wrong. (2) The only way of discovering the limits of the possible is to venture a little way past them into the impossible. (3) Any sufficiently advanced technology is indistinguishable from magic.

6 Charles Stross talking at the 2007 TNG Technology Consulting open day in Munich, reproduced on his blog posting of May 14, 2007, called "Charlie's Diary," at www.antipope.org.

7 Von Neumann was a prodigious polymath who made fundamental contributions to mathematics, physics, and computer science. He was a pioneer of quantum mechanics and played a key role in the Manhattan Project. He applied his concepts to economics and game theory. No dry mathematician, he was noted for his ribald and off-color humor, and he was a menace behind the wheel, prone to reading while driving. He described one incident this way: "I was proceeding down the road. The trees on the right were passing me in orderly fashion at 60 miles per hour. Suddenly one of them stepped in my path."

8 Paul Davies, quoted by Michio Kaku in "Star Makers," *Cosmos* no. 7 (2006), p. 12.

9 SETI researchers are fond of quoting Carl Sagan's aphorism "Absence of evidence is not evidence of absence." It's taken from the "Baloney Detection Kit" for detecting false and pseudoscientific arguments. He lists it as a fallacy of logic and rhetoric in his book *The Demon-Haunted World*. Non-detection of aliens is just that—a failure to detect them—and there are many possible explanations for a failed detection, only one of which is that they don't exist. There may be so many possible explanations that we have insufficient imagination to come up with them all. None of these alternative hypotheses are tested by the silence or the absence of evidence. We'd need to design different and better experiments to test them.

Chapter 11. How the Universe Ends

1 Hubble was preeminent in his field, but his measurements of cosmic distance and expansion got off to a rocky start, mostly because he had not understood that there were two types of variable Cepheid star, and he'd often been measuring entire star clusters when he thought he was measuring single stars. Hubble's initial measurements made the Milky Way seem larger than almost any other galaxy, and they led to an expansion rate that implied an age of the universe of just 2 billion years. Since radioactive dating of rocks gave an age for Earth of 3 billion years, it was a big problem to have a universe younger than objects within it! Hubble overestimated the expansion rate by a factor of 7, but his reputation has never suffered from this early blunder.

2 Cosmology makes a strong assumption that the universe is homogeneous and isotropic. Homogeneity is a statement that our part of the universe is no different, on average, from any other part. So the expansion we see is no different from the expansion seen by an observer on a distant galaxy. Testing homogeneity is quite difficult because we can't travel far enough to see for ourselves that all parts of the universe are similar, and if we look out in space we look back in time and so see distant regions as they were, not as they are. Isotropy means that the universe is no different looking in one direction than another. That assumption has been well tested by counts of distant galaxies looking in different directions away from Earth.

3 Helium-3, deuterium, and lithium were also produced in the big bang and their cosmic abundance matches predictions from the big bang model very well. The universe did not produce elements heavier than helium because as it expanded it rapidly became too cool for fusion to occur.

4 Stephen Hawking said that the theory of gravity contained the seeds of its own destruction in the form of black holes, because they embed a singularity of infinite temperature and density. The big bang makes a similar problem for cosmology because the expansion projects back to a state of infinite temperature and density, and the general relativity that describes the expansion cannot be used to calculate the initial state.

5 Special relativity puts a limit of the speed of light on the transmission of any signal between proximate parts of the universe, something that physicists call an inertial frame. But the universe is governed by the general theory of relativity, which has no speed limit. Two parts of the universe can be moving away from each other at more than light speed and, as a result, photons sent from those two places toward the middle will actually recede.

6 It's more accurate to think of the big bang as bounded in time rather than space. We can only see regions where the light has had time to reach us in the history of the universe since the big bang. Nothing rules out regions of space beyond the current light grasp of our telescopes.

7 Measurements of galaxies and quasars and the microwave background are sensitive to local geometry; the global geometry might not be measurable so it's important to have a theoretical motivation of what to look for. To keep things from getting too difficult to calculate, mathematicians tend to assume that the universe is a geodesically complete manifold, or a space where any two points can be connected by a shortest path. Space with ruptures called topological defects might not be amenable to calculation. Cylinders, soccer balls, and horns are not arbitrary shapes. They have a high degree of symmetry and correspond to fundamental positively and negatively curved spaces in three dimensions.

8 Normal particles—protons and neutrons—are called baryons, from the Greek for "heavy." The census of baryons is predicted by the big bang theory, but only 10 percent of the predicted numbers are found in all of the 50 billion galaxies and all their stars in the observable universe. Most of the rest are likely to be in a hot, diffuse gas in the vast spaces between galaxies.

9 In the case of dark energy, the biggest fear was a systematic error in the use of supernovae as standard lightbulbs. Because distant supernovae appear

unexpectedly faint, it was possible that dust in the wide spaces between galaxies was dimming them, but careful observations show that the supernovae didn't have the extra reddening expected from dust. Another possibility was a change over time in the supernova energy mechanism, perhaps because gas is being steadily enriched with heavy elements. But nobody has seen any sign of odd behavior in the supernovae, so they continue to be used as the best available tool to map out cosmic expansion.

10 M. Busha, A. Evrard, and F. Adams, "The Asymptotic Form of Cosmic Structures: Small-Scale Power and Accretion History," *The Astrophysical Journal* 665 (2007), p. 1.

11 Cosmologists have noted with interest the fact that there is a timing coincidence in the relatively recent transition from a decelerating to an accelerating universe. It happened relatively recently, less than half of the age of the universe ago. The corollary is that dark energy and dark matter are similar in magnitude and influence to within a factor of 3, which is close given their different basis in fundamental physics. If dark energy had been much larger, the universe would have expanded so fast that no structure would have formed, therefore biology could not have been possible either.

12 R. Caldwell, M. Kamionkowski, and N. Weinberg, "Phantom Energy: Dark Energy with $W < -1$ Causes a Cosmic Doomsday," *Physical Review Letters* 91 (2003), p. 71301.

13 Interview in *The New York Times* by Dennis Overbye, December 16, 2008.

14 Black hole evaporation has not yet been observed. All the black holes known, from the star-sized to the beasts that live in the centers of galaxies, are surrounded by hot gas and stars and many of them accelerate matter to phenomenal energies. As a result of all this activity, the subtle signature of Hawking radiation would be almost impossible to detect unless we were close to the event horizon.

Chapter 12. Beyond Endings

1 The epistemological status of surprise is difficult to ascertain. We might be surprised by some feature of the natural world because our theories and explanations are flawed and have led us in the wrong direction. We might be surprised because we had not understood all the natural consequences of successful theories that we had. We might be surprised because our theo-

ries encompassed phenomena that were far outside the realm of normal experience. Surprise doesn't necessarily imply a profound and deeper meaning, or point to an entirely new theory.

2 Michael Frayn, *The Human Touch* (New York: Henry Holt, 2006).

3 J. Barrow, F. Tipler, and J. Wheeler, *The Cosmological Anthropic Principle* (Oxford, England: Oxford University Press, 1988).

4 Steven Weinberg, quoted in Gian Giudice, "Naturally Speaking: The Naturalness Criterion and Physics at the LHC," arXiv:0801.2562.

5 Martin Rees, *Before the Beginning: Our Universe and Others* (New York: Simon and Schuster, 1997).

6 Steven Weinberg, in his opening talk at the symposium "Expectations of a Final Theory," held at the University of Cambridge in 2005 and published in *Universe or Multiverse?* ed. B. Carr (Cambridge, England: Cambridge University Press, 2006). However, skepticism of the correctness and predictive power of string theory and the multiverse idea has been growing, as exemplified in Peter Woit, *Not Even Wrong: The Failure of String Theory and the Search for Unity in Physical Law* (New York: Basic Books, 2006). The growing tide amounts to a backlash against the giddy optimism of the string theorists.

7 From a transcription of the question and answer session of the "Universe or Multiverse" presentations at Stanford University on March 26, 2003, as edited by Peter Chou on WisdomPortal.com.

8 Freeman Dyson, "Time Without End: Physics and Biology in an Open Universe," *Reviews of Modern Physics* 51 (1979), p. 447.

9 P. Steinhardt and N. Turok, "A Cyclic Model of the Universe," *Science* 296 (2002), p. 1436.

10 S. Lloyd, "The Ultimate Physical Limits to Computation," *Nature* 406 (2000), p. 1047.

Reading List

Numbers in parentheses refer to corresponding chapters in this book.

Adams, Fred, and Greg Laughlin. *The Five Ages of the Universe: Inside the Physics of Eternity*. New York: Free Press, 1997. (9)

Barrow, John, and Frank Tipler. *The Anthropic Cosmological Principle*. Oxford: Oxford University Press, 1986. (12)

Beech, Martin. *Terraforming: The Creating of Habitable Worlds*. New York: Springer, 2009. (8)

Benton, Michael. *When Life Nearly Died: The Greatest Mass Extinction of All Time*. London: Thames and Hudson, 2005. (6)

Berinstein, Paula. *Making Space Happen: Private Space Ventures and the Visionaries Behind Them*. Medford, NJ: Plexus, 2002. (8)

Blackmore, Susan. *Dying to Live*. Buffalo, NY: Prometheus, 1993. (1)

Boyd, Robert, and Joan Silk. *How Humans Evolved*. 4th ed. New York: W. W. Norton, 2005. (3)

Buettner, Dan. *The Blue Zones: Lessons for Living Longer from the People Who've Lived the Longest*. Washington, DC: National Geographic, 2008. (1)

Carr, Bernard, ed. *Universe or Multiverse?* Cambridge, England: Cambridge University Press, 2007. (12)

Carroll, Sean. *Endless Forms Most Beautiful: The New Science of Evo-Devo and the Making of the Animal Kingdom*. New York: W. W. Norton, 2005. (3)

Casoli, Fabienne, and Thérèse Encrenaz. *The New Worlds: Extrasolar Planets.* Berlin: Springer, 2005. (7)

Cheeta. *Me Cheeta: The Autobiography.* London: Fourth Estate Ltd., 2008. (1)

Cohen, Jack, and Ian Stewart. *What Does a Martian Look Like?: The Science of Extraterrestrial Life.* Hoboken, NJ: Wiley, 2002. (8)

Collins, Harry. *Gravity's Shadow: The Search for Gravitational Waves.* Chicago: University of Chicago Press, 2004. (10)

Dawkins, Richard. *The Ancestor's Tale: A Pilgrimage to the Dawn of Evolution.* Boston: Mariner Books, 2005. (3)

Diamond, Jared. *The Third Chimpanzee: The Evolution and Future of the Human Animal.* New York: Harper Perennial, 2006. (4)

Dixon, Dougal. *The Future Is Wild.* Buffalo, NY: Firefly, 2003. (4)

Dow, Kirstin, and Thomas Downing. *The Atlas of Climate Change: Mapping the World's Greatest Challenge.* Berkeley, CA: University of California Press, 2007. (5)

Enright, D. J., ed. *The Oxford Book of Death.* Oxford: Oxford University Press, 1983. (2)

Ferris, Timothy. *Coming of Age in the Milky Way.* New York: Harper Perennial, 2003. (9)

Finch, Caleb. *The Biology of Human Longevity.* Burlington, MA: Elsevier, 2007. (2)

Garreau, Joel. *Radical Evolution: The Promise and Peril of Enhancing Our Minds, Our Bodies—and What It Means to Be Human.* New York: Broadway, 2006. (4)

Golub, Leon, and Jay Pasachoff. *Nearest Star: The Surprising Science of Our Sun.* Cambridge, MA: Harvard University Press, 2002. (6)

Greene, Brian. *The Elegant Universe: Superstrings, Hidden Dimensions, and the Quest for the Ultimate Theory.* New York: W. W. Norton, 1999. (12)

Hawking, Stephen. *Illustrated Theory of Everything: The History and Fate of the Universe.* Beverley Hills, CA: Phoenix Books, 2009. (11)

Hazen, Robert. *Genesis: The Scientific Quest for Life's Origins.* Washington, DC: Joseph Henry Press, 2005. (3)

Kaku, Michio. *Parallel Worlds: A Journey Through Creation, Higher Dimensions, and the Future of the Cosmos.* New York: Doubleday, 2004. (12)

Kearl, Michael. *Endings: A Sociology of Death and Dying.* Oxford: Oxford University Press, 1989. (1)

Kurzweil, Ray. *The Singularity Is Near: When Humans Transcend Biology.* New York: Penguin, 2006. (4)

Largo, Michael. *Final Exits: The Illustrated Encyclopedia of How We Die.* New York: HarperCollins, 2006. (1)

Levy, David. *Shoemaker by Levy: The Man Who Made an Impact.* Princeton, NJ: Princeton University Press, 2002. (7)

Lovelock, James. *The Revenge of Gaia: Earth's Climate Crisis and the Fate of Humanity.* New York: Basic Books, 2006. (5)

Lunine, Jonathan, and Cynthia Lunine. *Earth: Evolution of a Habitable World.* Cambridge, England: Cambridge University Press, 1998. (5)

Mayr, Erst. *What Makes Biology Unique?* Cambridge, England: Cambridge University Press, 2004. (3)

Melia, Fulvio. *The Edge of Infinity: Supermassive Black Holes in the Universe.* Cambridge, England: Cambridge University Press, 2003. (9)

Mitchell, Stephen. *Gilgamesh: A New English Version.* New York: Free Press, 2006. (2)

Plait, Phillip. *Death from the Skies!: These Are the Ways the World Will End.* New York: Penguin, 2008. (7)

Posner, Richard. *Catastrophe: Risk and Response.* Oxford: Oxford University Press, 2005. (6)

Preston, Richard. *The Hot Zone: A Terrifyingly True Story.* New York: Anchor, 1999. (6)

Rees, Martin. *Our Final Hour: A Scientist's Warning.* New York: Basic Books, 2003. (6)

Rees, Martin. *Just Six Numbers: The Forces That Shape Our Universe.* New York: Basic Books, 2000. (11)

Schmidt, Stanley, and Robert Zubrin, eds. *Islands in the Sky: Bold New Ideas for Colonizing Space.* New York: Bantam Doubleday, 1996. (10)

Schrödinger, Edwin, and Roger Penrose. *What Is Life?* Cambridge, England: Cambridge University Press, 1992. (3)

Singh, Simon. *Big Bang.* New York: HarperCollins, 2004. (11)

Steinhardt, Paul, and Neil Turok. *Endless Universe: Beyond the Big Bang.* New York: Doubleday, 2007. (11)

Taylor, Travis, and Bob Boan. *An Introduction to Planetary Defense: A Study of Modern Warfare Applied to Extraterrestrial Invasion.* Boca Raton, FL: Brown Walker, 2006. (10)

Tipler, Frank. *The Physics of Immortality: Modern Cosmology, God, and the Resurrection of the Dead*. New York: Random House, 1994. (2)

Waller, William, and Paul Hodge. *Galaxies and the Cosmic Frontier*. Cambridge, MA: Harvard University Press, 2003. (9)

Ward, Peter. *Under a Green Sky: Global Warming, the Mass Extinctions of the Past, and What They Can Tell Us About Our Future*. New York: Collins, 2008. (6)

Ward, Peter, and Donald Brownlee. *The Life and Death of Planet Earth: How the New Science of Astrobiology Charts the Ultimate Fate of Our World*. New York: Henry Holt, 2002. (5)

Ward, Peter, and Donald Brownlee. *Rare Earth: Why Complex Life Is Uncommon in the Universe*. New York: Springer, 2000. (7)

Webb, Stephen. *If the Universe Is Teeming With Aliens . . . Where Is Everybody? Fifty Solutions to Fermi's Paradox and the Problem of Extraterrestrial Life*. New York: Springer, 2002. (10)

Wharton, David. *Life at the Limits: Organisms in Extreme Environments*. Cambridge, England: Cambridge University Press, 2007. (4)

Young, Simon. *Designer Evolution: A Transhumanist Manifesto*. Amherst, NY: Prometheus, 2006. (8)

Zubrin, Robert. *Entering Space: Creating a Spacefaring Civilization*. New York: Tarcher, 2000. (8)

Credits

Figure 1.1 Crawford Collection, Royal Observatory Edinburgh. *Figure 1.2* L. Rowell Huesmann, University of Michigan. *Figure 1.3* Joao Pedro Magalhaes, University of Liverpool. *Figure 1.4* Knut Schmidt-Nielsen, Duke University, and Cambridge University Press. *Figure 1.5* Eurostat, European Communities. *Figure 1.6* René Descartes, Metaphysical Meditations (1641). *Figure 2.1* Bjorn Lomborg, Aarhus University, and Cambridge University Press. *Figure 2.2* Centers for Disease Control and Prevention, National Center for Health Statistics. *Figure 2.3* T. O'Donnell, History of Life Insurance in Its Early Years (1916). *Figure 2.4* Carol Vleck, Iowa State University, and Springer. *Figure 2.5* Marek Rowland-Mieszkowski, Digital Recordings. *Figure 2.6* National Oceanographic and Atmospheric Administration Earth System Research Lab and the Intergovernmental Panel for Climate Change. *Figure 3.1* Randolph Femmer, National Biological Information Infrastructure and the U.S. Geological Survey. *Figure 3.2* Sanba38, Wikipedia GNU License and Creative Commons License. *Figure 3.3* Kazuhiko Ohshima, Tokyo Institute of Technology. *Figure 3.4* Albert Mestre, Wikipedia GNU License and Creative Commons License. *Figure 3.5* GRIDA/UNEP and the Intergovernmental Panel for Climate Change. *Figure 4.1* TedE, Wikipedia GNU License and Creative Commons License. *Figure 4.2* National Institutes of Health. *Figure 4.3* U.S. Geological Survey, Department of the Interior. *Figure 4.4* Samantha, and Wikipedia Creative Commons. *Figure 4.5* Adapted from an original concept by Frank Drake, UC Santa Cruz. *Figure 5.1* John Valley, University of Wisconsin–Madison. *Figure 5.2* Robert Hazen and David Deamer, NASA Astrobiology Institute and Carnegie Institution of Washington. *Figure* 5.3 NASA Astrobiology Institute and the Indiana-Princeton-Tennessee Team. *Figure* 5.4 Hugo Alhenius, UNEP/GRID-Arendal. *Figure* 5.5 NASA/Jet Propulsion Laboratory. *Figure* 6.1 Chris Chapman, HoneyBee Robotics, and David Morrison, NASA Ames Research Center. *Figure* 6.2 NASA Headquarters and the Crew of STS-9. *Figure* 6.3 Howard Lester, Multiple Mirror Telescope Observatory, and the Large Synoptic Survey Telescope Corporation. *Figure* 6.4 Pat Rawlings, University of Maryland, and NASA/Jet Propulsion Laboratory. *Figure* 6.5 NASA/Johnson Space Center. *Figure* 6.6 Anita Heward, Europlanet, and ESA. *Figure* 7.1 John Valley, University of Madison–Wisconsin. *Figure* 7.2 Geoff Marcy, UC Berkeley, and the California and Carnegie Planet Search. *Figure* 7.3 Nick Strobel, UC Bakersfield, and Astronomy Notes. *Figure* 7.4 Justin Cantrell, Todd Henry, Georgia State University, and the RECONS Survey. *Figure* 7.5 Don Yeomans, NASA/Jet Propulsion Laboratory. *Figure* 7.6 Klaus Knie and Gunther Korshinek, Munich Technical University. *Figure* 8.1 Robert A. Rohde, UC Berkeley, and Global Warming Art. *Figure* 8.2 Tablizer, Wikipedia GNU License and Creative Commons License. *Figure* 8.3 NASA/Glenn Research Center. *Figure* 8.4 Natasha Vita-More, Extropy Institute. *Figure* 8.5 Danila Medvedev, Russian Transhumanist Movement. *Figure* 9.1 Steward Observatory and Kitt Peak National Observatory/National Optical Astronomical Observatories. *Figure* 9.2 Carlos Frenk, University of Durham, and Millennium Simulation/VIRGO Consortium. *Figure* 9.3 NASA/Chandra X-Ray Observatory. *Figure* 9.4 John Dubinski, University of Toronto. *Figure* 9.5 Reinhard Genzel and Thomas Ott, Max Planck Institute for Extraterrestrial Physics. *Figure* 9.6 Alexander J. Blustin, University of Cambridge, Matthew Page and Rebecca Smith, UCL/Mullard Space Science Laboratory. *Figure* 10.1 NASA/Jet Propulsion Laboratory. *Figure* 10.2 NASA/Space Telescope Science Institute. *Figure* 10.3 Chris Henze, NASA/Ames Research Center, Joan Centrella, NASA/Goddard Space Flight Center, and the NASA/Advanced Supercomputing Division. *Figure* 10.4 Chris Impey. *Figure* 10.5 NASA/Ames Research Center. *Figure* 11.1 Robert Kirshner, Harvard University, and the Proceedings of the National Academies of Sciences. *Figure* 11.2 NASA/Wilkinson Microwave Anisotropy Probe Science Team. *Figure* 11.3 Kirk Korista, Western Michigan University. *Figure* 11.4 Hobby-Eberly Dark Energy Experiment, McDonald Observatory. *Figure* 11.5 David Caldwell, UC Santa Barbara, and the American Institute of Physics. *Figure* 11.6 Chris Impey. *Figure* 12.1 Chris Impey. *Figure* 12.2 Chris Impey, based on a concept by Paul Davies, Arizona State University. *Figure* 12.3 Chris Impey. *Figure* 12.4 Scott Aaronson, Massachusetts Institute of Technology. *Figure* 12.5 Chris Impey and Linden Labs/Second Life. *Figure* 12.6 Chris Impey.

Index

Page numbers in *italics* refer to illustrations.
Page numbers beginning with 301 refer to notes.

About Schmidt, 51
accidents, 19, 20, 22–23, 46, 48–50,
    194, 286
Ackerman, Diane, 217
actuaries, 23, 51, 291
Adams, Douglas, 180
Adams, Fred, 198–99, 232–33, 237,
    268, 270
Addison's disease, 145
aerosol, 82, 87, 89
Africa, 75, 76, 83, 94, 203
afterlife, 23, 33–34, 36, 38–39, 287
Agamemnon, 18–19, 20
aging, 52–56
    accelerated, 53
    of cells, 34–35, 52, 54
    disease and, 52, 54, 55
    evolutionary process and, 53
    fertility and, 55
    fitness and, 54, 55
    free radicals and, 52–53
    genes and, 30, 30, 33, 53, 55–56
    inexorability of, 52–53

interplay of factors in, 53–56, 54
    slowing of, 52, 55, 56, 104–6, 123,
        199
    theories of, 52–53, 55
Agricultural Revolution, 26
Air Force, U.S., 139
Alcor Life Extension Foundation, 59,
    200
algae, 124, 125, 127, 133
Allan Hills meteorite 84001, 154, 155
alleles, 73–74, 101–2, 291
Allen, Woody, 22
Almagest, The (Ptolemy), 208
Alpha Centauri, 242, 271
Alzheimer's disease, 57
American Geophysical Union, 85–86
American Medical Association, 49
American Petroleum Institute, 182
Ames Research Center, 196
amino acids, 116, 118, 307
ammonia, 116, 118, 169, 197
amphibians, 28, 29
Anaxagoras, 207, 317

Andromeda galaxy (M31), 218–21, 220,
    223–26, 229, 235, 237, 250,
    266, 291, 319
anesthetic, 40, 44
Animal Ageing and Longevity Database,
    29
animals, 26–33, 55–56, 73, 112, 151,
    202, 247
    correlation between size and longevity
        of, 28–30, 29
    correlation between size and meta-
        bolic rate of, 29–31, 30
    diseases of, 66
    interbreeding of, 71
    oldest living, 28
    research on, 88
    see also specific animals
Ansari X Prize, 193
Antarctica, 112, 119, 168, 187, 310
anthrax, 87, 88
anthropic principle, 272, 275–77, 280,
    291
antiaging, 55
    research in, 52, 56, 104–6
Anti-Ballistic Missile Treaty of 1972, 84
antibiotics, 88, 97, 99
antigravity, 362
antioxidants, 53
anti-Semitism, 42
ants, 43, 205, 290
apes, 12, 31–33, 71, 76, 135, 238,
    306–7
Apollo missions, 80, 81, 145, 243
Archaea, 120, 122, 123, 156, 291, 310
Argiope appensa, 69
Aristotle, 25, 36, 115, 208, 317
Arizona, University of, 145, 148
Arizona State University, 280
Armageddon, 145
Arrhenius, Svante, 153
art, 19, 21–22, 31, 67
ASPM gene, 101
asteroids, 113, 115, 123, 134–51, 179,
    197, 244
    avoidance of, 146–49, 150
    Earth-crossing, 114, 135–36, 144,
        146, 172, 293, 312

asthma, 43, 83
astrobiology, 119, 190
astrogeology, 145
astronauts, 149, 184, 192–93
astronomers, 51, 106–7, 131, 137, 144,
    153, 160, 163, 168, 175, 181–
    83, 190, 195, 208–26, 241, 250,
    257, 279–80, 288–89
astrophysics, 160–61
    computational, 212–15, 292
As You Like It (Shakespeare), 212
Atacama Desert, 119–20, 189, 209, 309
atheism, 25
atmosphere, 82, 114
    carbon in, 63, 63, 77, 80, 83
    cooling of, 85–86, 94–95, 127
    smoke and soot in, 85–86, 135
    testing of, 111
    see also climate; climate change;
        Earth, atmosphere of; global
        warming; ozone cover
Atomic Energy Lab (U.K.), 194
atoms, 11, 38, 52, 56, 195, 207–8, 211,
    217
    carbon, 62–64, 227–28
    shredding of, 266, 267
Attila the Hun, 23
Aum Shinrikyo terrorist group, 86
Australia, 28, 95, 105, 113, 114, 115,
    145, 309
Australian aborigines, 77
autopsy, 45

bacteria, 77, 96–98, 97, 151, 152, 155–
    56, 310
    anaerobic, 92
    antibiotic-resistant, 97, 99
    colonies of, 26, 92
    genetics of, 98, 121–22
    species of, 93, 112, 121–23, 122, 132,
        152
    weaponized, 87–88
bacterium SAR-11, 77
Bad Astronomy blog, 179
Bailey, Ronald, 201–2
Baliunas, Sallie, 181–82
Barker, Lex, 31

Barnard's Star, 174
Barrett, Syd, 191
Barringer crater, 139, 142, 144–45
Barrow, John, 276
basalt, 152, *155*
Beagle, 28
Beatles, 28
Becker, Lance, 61
beetles, 90, 92, 93, *97,* 205
Bela Lugosi Meets a Brooklyn Gorilla, 31
Bell Labs, 222, 253
Bennett, Charles, 195, 315
Berlin Wall, 84
Berra, Yogi, 238
beryllium-10, 186
Beyond Humanity (Paul and Cox), 108
Bible, 102
big bang, *214,* 217, 227, 229, 234, 252–
    54, *253,* 256–57, 275, 282, 290,
    292, 321–22
  microwave radiation from, 26, 253,
    257–58, 264, 322
Big Dipper, 205
big rip, 266, *267,* 292
biodiversity, 78–79, *79,* 93–94, 112, 151,
    158, 292
biological sensing equipment, 123–24
biological weapons, 84, 86–89
Biological Weapons Convention of 1972,
    86, 78
biology, 52, 56, 66–75, 112, 114, 124–
    25, 131, 149, 201, *201,* 271–72,
    273, 275, 277
  speculative, 90–91, 284
  transcending limits of, 12, 35, 102–9
Biopreparat, 87
biosphere, 38, 57, *57,* 62–63, *63,* 114–
    15, 188–89, 292
  shadow, 121–24
  threats to, 12, 134–56, 311–12
biotechnology, 152, *201*
bird flu, 98, 99
birds, 28, 29, 91, 92, 115, 134
birth defects, 60
Black Cloud, The (Hoyle), 281
Black Death, 44
black holes, 177–79, 185, 211, 216,

217, 218, 222–26, *224, 226,*
    230, 234, 235, 237, 248, 251,
    265, 269–70, *269,* 284–86, *285,*
    290, 292, 299, 314, 319, 323
blood, 11, 17, 34, 43, 44, 61
bodies, 34–39
  aging and renewal of cells in, 34–35
  burial of, 43, 50
  cooling of, 17, 61–62
  corpses of, 22, 38, 39, 66
  deterioration of, 52–54, 58
  development of, 53
  dismembering and eating of, 66
  freezing of, 58–60, 62, 200, 293, 305
  immobilized, 40
  minds vs., 36–38, *37*
  rebirth of, 34, 38–39
  repair of, 55, 58, 104
  vestigial attributes in, 74
  weight of, 62
Bohr, Niels, 238
Bond, Robert, 194
bones, 34, 43, 104, 124, 215
Bostrom, Nick, 202, 271–72, 286
Boswell, James, 25
botany, 67–68
botulism, 87
Bowyer, Adrian, 244
brains, 17–19, 58, 76, 159, 283, 284,
    303
  age-related changes in, 57
  altered functioning of, 18
  correlation between life span and
    weight of, 29
  deterioration of, 19, 40, 58, 80, 88, 89
  electro-chemical connections in, 18,
    40, 59, 60–61, 87, 202, 283
  extinction of activity in, 18, 39, 60
  freezing of, 202
  growth and development of, 101, 135
  loss of oxygen in, 18, 60–62
  minds vs., 18, 36, *37*
  plasticity of, 77
  size of, 76, 77, 135
Branson, Richard, 194
Brilliant, Larry, 98–99
bristlecone pines, 27, 55

Bronze Age, 141
Brownlee, Don, 159, 189
B612 Foundation, 149
bubonic plague, 44, 86
Buchwald, Art, 22
Buddhism, 39, 302
Bulletin of the Atomic Scientists, 84–85
Bush, George W., 93
Busha, Michael, 264
butterflies, 69, 72

cadmium, 80, 121
Caenorhabditis elegans, 97
calcium, 125, 215
Caldwell, Robert, 266
California, University of, 307
    at Berkeley, 173
    at Los Angeles, 200, 222
    at Santa Cruz, 118
California Institute of Technology
    (Caltech), 222, 285
Callisto, 168, 172
Calment, Jeanne Louise, 32–33
Cambrian explosion, 77, 133, 151
Cambridge University, 280, 282
cancer, 24
    death from, 46, 47–48, 49, 52
    risk factors for, 54, 80, 81, 95
    treatment and prevention of, 47, 61
cannibalism, 66
carbohydrates, 62, 125
carbon, 62–64, 125–26, 160, 175, 191,
    215, 227–28, 259, 269
    in atmosphere, 63, 63, 77, 80, 83
    cycling of, 62–63, 63, 83, 292
    radioactive, 116
carbon dioxide (CO_2), 64, 77, 80, 82,
    83, 120, 125, 127, 132–33, 132,
    152, 166, 188–89, 195
    human activity and increasing levels
    of, 126
    recycling of, 167
carbon monoxide (CO), 64, 118
carbon offsetting, 129
cardiac arrest, 17–18, 40, 60–62
cardiopulmonary resuscitation (CPR), 60
Carnegie Institute, 118

Carnegie-Mellon University, 283
Carol Burnett Show, The, 59
Carroll, Lewis, 98
Carson, Rachel, 79–80, 81, 111
Carter, Brandon, 165, 275, 313
Carter, Jimmy, 81
Cassini orbiter, 169, 198
Cassiopeia constellation, 206
Castle Bravo device, 139
cataracts, 52, 53
Catastrophe Calculator, 145
cats, 135, 164
caves, 120
Celestis, Inc., 64
cells, 37, 52–53, 55, 96, 114, 283
    aging and replacement of, 34–35, 42,
        52–54
    damage and death of, 52–53, 54, 59,
        61–62
    division of, 26, 53–54, 100, 123
    heart, 61, 62
    prokaryotic, 117, 123, 128, 132, 310
centenarians, 32–33, 33, 54, 292, 302
cerebral cortex, 35
Challenger explosion, 19
chaotic inflation theory, 279
Cheeta (chimp), 31–32
chemicals, 114, 116, 158
    toxic, 80–81, 86–87, 118–19, 120
    weaponized, 86–87
chemistry, 153, 273
Chicago, University of, 101, 245
childbirth, 42, 45, 46, 50, 60, 100
children, 43, 60, 100
    mortality of, 46, 48
chimpanzees, 31–32, 71, 307
    divergence of humans and, 76–77, 95,
        97, 101
China, 84, 193, 197
chlorophyll, 133
cholera, 46, 83, 87, 239
Christianity, 38, 113, 200
chromosomes, 53–54, 68, 70, 71, 94
Churchill, Winston, 30
cigarette smoking, 32, 49, 50
civilizations, 101–2, 160, 207, 286–87
    alien, 106–8, 107, 109, 112

communication between, 106–8, *107*, 109

clams, 27, 55

Clean Air Act of 1970, 81

Cleland, Carol, 123, 311

climate, 78–79, 126–30
feedback cycles and, 127–28, *128*, 129–30
impact of nuclear war on, 85–86

climate change, 81–83, *82*, *187*, 292
effects of the Sun on, 181, 182, 183, 185–87
human activity and, 85–86, 126, 182, 186
see also global warming

Clive, Robert, 28

clonal colonies, 27, 292, 302

cloning, 106

cloud cover, 127, *128*

Clovis culture, 140

coelacanth, 94

Cohen, Jack, 165

Cohen, Marc, 35

cold war, 84, 139, 193

Colorado, University of, 123

Columbia space shuttle, *143*

coma, 17–18

Comet 1862 III, 141

Comet Encke, 141

comets, 136–37, 139–42, 146–47, *150*, 151, 153, 172–75, *173*, 184, 197

Comet Shoemaker-Levy 9, 144

Comet Tempel 1, *150*

Commercial Space Launch Act of 1984, 193

computer technology, 26, 40, 85, 88, 98–100, 103–5, 107, 145, 147, 185, 202, 218–19, 238, 246, 271, 284
simulation programs with, 212–15, *214*, *220*, 235, 286–87, 318

conservationism, 94

Contact, 51

Cook, James, 28

Copernicus, Nicklaus, 160–61, 165, 250, 275, 279, 287, 289, 293, 313

copper, 124, 215

cosmic rays, 175, 186, 241

Cosmological Anthropic Principle, The (Barrow and Tipler), 276

cosmos, 12, 121, 226, 233, 289
geocentric model of, 208
heliocentric model of, 254

Cox, Earl, 106

Cox, T. J., 221

creationism, 114, 277

cremation, 59, 64

creosote bush, 27

crocodiles, 92, 115

crows, 70–71

cryonics, 58–60, 62, 200, 293, 305

Cryonics Institute, 59

Cueva de Villa Luz, 120

Curry, Oliver, 101

cyanobacteria, 117, 130, 132–33

cyborgs, 104–5, 239, 240, 293, 316

cyclic universe theory, 282, 293

Dalai Lama, 23–24

dark energy, 260–68, *261*, 270, 282, 293, 322–23

dark matter, 211, 212–14, *214*, 224, 237, 260, *261*, 262, 263, 265, 275, 293, 317, 323

Dartmouth College, 266

Darwin, Charles, 28, 53, 70, 72, 100, 116, 117, 118, 153

Davies, Paul, 244–45, 280, 321

Davis, Devra, 81

Deamer, David, 118

Dean, James, 19

death, 12, 19–26, 158–59, 188, 303–5
accidental, 19, 20, 22–23, 46, 49–50, 286
artistic depictions of, 21–22
belief in life after, 23, 33–34, 36, 38–39, 287
boundary between life and, 39–40
causes of, 22–25, *24*, 42, 43, 44–50, 47
continuum of life and, 21–22
debate or definition of, 60–61
enigma of, 22, 25

death (*continued*)
  fear of, 30, 51, 287
  fiction and nonfiction accounts of,
    21–25
  forestalling of, 58, 199–202
  inevitability of, 19–21, 33, 35, 48
  infant, 43, 45–46
  maternal, 45, 46, 50
  natural process of, 21–22
  obsession with, 51
  predictability of, 42, 50–51
  reminders of, 21–22
  seeing humor in, 22, 23
  study of, 23–25
  sudden, 19, 20, 23, 25
  suffering and, 30, 66
  timing and patterns of, 50–51
  Western cultural aversion to, 21, 22,
    23
  see also disease, terminal; suicide
Death by Black Hole (Tyson), 177
death certificates, 24
*Death from the Skies: These Are the
  Ways the World Will End* (Plait),
  179–80
death rates, 46–50, 47, 304
Deep Impact (film), 145
Deep Impact space probe, 150
defibrillators, 61
de Magalhães, João Pedro, 29, 30
Demetrius Phalereus, 34
Democritus, 207–8, 317
de Moivre, Abraham, 50–51
Dengue fever, 87, 99
Descartes, René, 36, 37, 37, 303
deserts, 112, 158
Desulforudis audaxviator, 122–23, 122,
  189
diabetes, 44, 103
diamonds, 140, 191–92
diarrhea, 48
diatoms, 68
Dick Cavett Show, The, 23
diet, 33, 44, 45, 48, 103
dinosaurs, 134, 135, 140, 146
Discover, 179, 310
disease, 21, 43–46, 47, 97

aging and, 52, 54, 55
animal, 66
control and eradication of, 98, 104,
  199, 239
degenerative, 44, 52
epidemic and pandemic, 65, 98, 99,
  101, 153, 297
genetic, 95
prevention of, 47, 49
risk factors for, 46, 54, 80, 81, 95
terminal, 20, 22–25, 24, 42, 43,
  44–48, 47, 52, 65–67, 99, 286
transmission of, 44, 45, 46, 47, 83,
  87, 99
Disney, Walt, 22, 59
disposable soma, 53
Dixon, Dougal, 90–91
DNA, 23, 54, 66–67, 71, 72–73, 88,
  94–95, 97, 101, 112, 116, 117,
  123, 217, 307, 310
Doctor Doolittle, 31
dogs, 30, 50, 142, 161, 242, 280
Doomsday Clock, 84–85, 108–9, 313
Doppler effect, 164, 313
Drake, Frank, 106–8
Drake equation, 106–8, 247, 290, 293,
  309
Drexler, Eric, 244
driving, 46, 48–49, 50, 100
drugs, 81, 100, 104, 191
Druyan, Ann, 25
dualism, 35–38, 52
Duane, Don, 23
Dubinski, John, 218–21, 220
Durham University Institute for Compu-
  tational Cosmology, 212
Dyson, Freeman, 280–82, 283, 324

Eagles, 18
Earth:
  atmosphere of, 26, 64, 79–83, 111–
    12, 121, 124, 125–30, 133, 138,
    153, 158, 222
  clones of, 240
  cooling of, 129–30, 133, 139–40, 150,
    181, 185–87
  crust of, 122, 125, 126

death of, 13, 159, 189–90, 192, 195, 198, 266
Epoch Zero through Epoch Five on, 132–33
equatorial land mass of, 129
formation of, 26, 113–14, 124
fragile ecosystem of, 80
history and evolution of, 11, 26, 113–18, *161*
impact of meteors and asteroids on, 113, 114, *115*, 116–17, 123, 132, 134–51, *138*, *143*, *150*, 157–58, 172, 173, 187, 311–12
magnetic field of, 241
orbit of, 64, 136, 141, 142, 144, 172, 187, 240, 242, 275, 279, 311, 315
periods of mass extinction on, 78–79, *79*, 135–40, *138*, 150, 151–52, 173, 179, 296
rotation of, 289, 311
as seen from afar, 80, 111, 131–33, *132*, 136
survival and future of, 90–91, 137–56, 158–65
survival of humans outside of, 12, 152, 165, 195–98, 239–43, 280–90
temperature of, 57, 126–27
threats to, 134–56, 172, 175–80, 311–12
tilting axis of, 187
earthquakes, 134
East India Company, 28
Ebola fever, 87, 88
ecosystems, 77, 79, 92
  microbial, 96–98, 121–23
eggs, 70, 75
Egypt, 67, 86
Einstein, Albert, 265, 266, 318
  $E=mc^2$ equation of, 84, 188, 215
  relativity theory of, 236, 250–51, 262–63
elasmosaurs, 134
electromagnetism, 165, 273, 279, 281, 320
electrons, 36, 38, 52, 268, 270, 278

elephants, 28, 29, 72, 108, 135
Elizabeth II, Queen of England, 28, 32, 279
embryos, 59–60, 75
Enceladus, 169
encephalomyelitis, 88
endoliths, 122
endorphins, 40
energy, 56–58, *57*, 61, 78, 116, 125, 168, 278
  chemical sources of, 121, 122–23, *122*
  conserving of, 281
  conversion of matter into, 274
  fixed state of, 282
  flow of, 56
  nuclear, 84, 106, 139, 145, 175
  phantom, 266, 267
  release of, 84, 139, 145, 178, 180, 269
  see also dark energy
entropy, 52, 56–58, *57*, 293, 304
environment, 39, 52, 55, 58, 73–74, 97, 286
  evolution and, 74, 78, 92, 94–95, 116–21, 135
  genes and, 73–74
  life span and, *30*, 55
  variations and extremes of, *54*, 116–21
environmental movement, 111
Environmental Protection Agency, U.S., 80
enzymes, 55, 101, 133
Epic of Gilgamesh, The, 102–3
epidemiology, 98–99, 153
Eridanus constellation, 203
Erlich, Robert, 174
Eta Carina, 180
Ettinger, David, 60
Eudoxus, 208
eukaryote, 128, 156, 293, 310
Europa, 168–69, 172, 191, 310
European Space Agency, 155–56, *155*, 259
evolution, 89–94, 124, 153
  acceleration of, 97–98

evolution (*continued*)
  aging and, 53
  changing environments and, 74, 78,
      92, 94–95, 116–21, 135
  Darwin's theory of, 53, 116, 117
  mutation and, 73–74, 92, 95, 101
  phylogenetic techniques and, 72–73
  random change and, 100, 101
  social, 76–77, 95
  of species, 11, 72–77, 90–94, 189
  three major components of, 100–101
  see also natural selection; *specific spe-*
      cies and systems
evolutionary developmental biology (evo-
      devo), 73, 294
execution, 49
exercise, 31, 49
exoplanets, 131, 133, 159, 160, 162–64,
      164, 169–70, 294, 312–13
extraterrestrials (E.T.s), 123, 133, 153–
      56, 161–62
extremophiles, 119–23, *122*, 154, 159,
      294, 316
ExxonMobil, 182
eyes, 34–35, 92

Farah, John, 221
Fautin, Daphne, 92
Federal Aviation Administration (FAA),
      193
Fermi, Enrico, 245–46, 294
Fertile Crescent, 141
fevers, 45, 87, 98, 99
Final Exits (Largo), 24–25
fire, 43, 49
First and Last Men (Stapledon), 195
Fischer, Bobby, 22
fish, 27–28, 29, 91, 94, 134, 280
fitness, 54, 55
Five Ages of the Universe, The (Adams
      and Laughlin), 233
Fixx, Jim, 23
FM-2030 (F. M. Esfandiary), 200
folklore, 142, 183
folk taxonomies, 68
food, 32, 43–44, 76

chain of, *68*, 134–35, 184
  nutrients in, 44, 56, 62
  threats to production of, 83, 86
forests, 205–6, 216
  loss of, 126, 139, 179
Forest Service, U.S., 27
Fore tribe, 65–66
formaldehyde (H_2CO), 84, 120
Fortey, Richard, 93, 308
fossil fuel, 125, 126
fossils, 71, 76, 78, 113, 123, 152, *155*,
      156, 176
Foster, Jodie, 51
Foton capsule, 155–56
Frayn, Michael, 276, 324
free radicals, 53, 294
free will, 287, 288
Frenk, Carlos, 212–15, 218
Frost, Robert, 265
Fukuyama, Francis, 201
fungi, 43, 72, 87, 310
future, 12
  of humanity, 65–91, 106, 137–65,
      199–202, 238–46, 280–90
  prediction of, 11, 90–91, 238, 289
  Utopian vs. dystopian vision of, 104–5
Future of Humanity Institute, 202
futurology, 238–40, *239*, 294

Gaia, 110–12, 124–29, 294
Gajdusek, Carleton, 66
galaxies, 11, 12, 106, 109, 192, 250–52,
      251, 255–56, 289
  dwarf, 221
  elliptical, 235
  evaporation of, 236, 294
  formation of, 211–15
  see also Milky Way; *specific galaxies*
Galileo Galilei, 168, 172, 208–9, 265
Galileo probe, 131, 198
gametes, 26
gamma rays, 175, 176
  bursts of, 178–80, 218, 269
Gamow, George, 252–53
Ganymede, 168, 172
Garreau, Joel, 104

gases, 112, 125, 129
   toxic, 86, 120, 152
   see also greenhouse gas emissions; *specific gases*
Gates, Bill, 238
Gaunt, John, 50
GD 362, 195
genes, 39, 72–75, 117, 156, 306
   aging and, 30, *30, 33,* 53, 55–56
   duplication of, 75, 117
   environment and, 73–74
   exchange of, 70, 71, 74–76
   horizontal transfer of, 97, 306
   information stored in, 66–67, 73–74
   mix and variation of, 101–2
   mutation of, 73–74, 92, 95, 101
   rogue, 53
   sequencing of, 88
gene therapy, 104, 238
genetic drift, 74, 95, *96*
genetic engineering, 88–89, 104–6, 152, 239
genetic revolution, 104
genetics, 95, 98, 100, 104–6, 201, 302
Geneva Protocol of 1925, 86
Genographic Project, 95
genomes, 88, 94, 95, 101–2, 283, 294
Genzel, Reinhard, 223
geochemistry, 151–52
geology, 113, 114, 124–25
germ theory, 45, 115
Ghez, Andrea, 222–23
Ghost, 35–36
ghosts, 35–36
Gilgamesh, 102–3, *103,* 141–42, 200
Gliese 710, 174
global geometry, 258–59
Global Oscillation Network Group, 184–85
global warming, 78–79, 81–83, *82,* 84, 295
   causes of, 126, 127, 129
   effects of, 83, 91, 183, 186
   human-induced, 126, 186
   possible irreversibility of, 129
glucose, 57, 61

God, 36, 38–39, 92, 277
Goddard, Robert, 240
Goldberg, Whoopi, 35
Golding, William, 126
Gompertz, Benjamin, 41–42, 50–51, *51*
Goodall, Jane, 31, 307
Google, 98–99, 308
Google X Prize (Moon 2.0), 194
Gorbachev, Mikhail, 85
gorillas, 31, 72, 76
Gott, Richard, 160–61, 165, 313
Gould, Stephen Jay, 277
Gravitas, 221
gravity, 114, 124, 154, 190, 191, 213, 251, 264, 273, 282, 322
   artificial, 243
   cosmic structure formed by, 259
   expansion of universe opposed by, 252, 263
   see also antigravity; *specific gravitational effects*
Gravity's Rainbow (Pynchon), 21, 143
gravity wave experiments, 282, 295
gravity waves, 282, 295
Great Dying, 78–79, 295
Greeks, ancient, 25, 34, 35, 50, 67, 113, 126, 203–4, 206–8, 254, 255
green goo, 152, 290, 295
greenhouse gas emissions, 82, *82,* 129, 130, 133, 153, *167*
Greyson, Bruce, 40
Griffin, Michael, 240
Grinspoon, David, 164, 166, 313
Guinness Book of World Records, 31
Gulf Stream, 186
Gulliver's Travels (Swift), 183

habitat, 78–79, 94
Hadean period, 113, 157, 295
Haldane, J. B. S., 92, 116, 201
Halley, Edmund, 50–51
Halley's Comet, 89, 141
hallucination, 18, 40
Hannah, Daryl, 51
Haraway, Donna, 200, 316
Harris, Alan, 144

Harrison, Rex, 31
Harvard-Smithsonian Center for Astro-
      physics, 221
Harvard University, 117, 261, 277
   Astronomy Department of, 224
Hawaii, University of, 131
Hawking, Stephen, 85, 194, 265, 269,
      269, 284–86, 285, 295, 315,
      322, 323
Hawks, John, 101–2
Hayflick limit, 53–54, 295
Hazen, Robert, 118
heart:
   restarting of, 18, 60, 61, 305
   stopping of, 17–18, 40, 60–62
heart attack, 23, 62
heartbeats, 12, 29, 31
heart disease, 46–47, 52, 103
heat, 57, 63, 80, 113
Heinlein, Robert, 30
helium, 64, 125, 188, 188, 191, 213,
      215, 253, 269, 273, 318
herbal remedies, 43
Herodotus, 185
Herschel, Caroline, 209
Herschel, Sir William, 183, 209–10,
      250, 254–55, 314–15
Hesiod, 126
hibernation, 28, 281
Hinduism, 39, 302
Hipparcos satellite, 174
Hippocrates, 207–8
Hippocratic Oath, 44, 303
Hirst, Damien, 22
Hitchhiker's Guide to the Galaxy, A
      (Adams), 180
HIV/AIDS, 48, 97, 98, 99, 101
Hizbullah, 22
Hobbes, Thomas, 43
Hodges, Ann, 142, 144
Holden, Jan, 40
Holocaust Museum, 59
Holocene Impact Working Group,
      139–40
Holocene period, 186
Homer, 19, 102
Homo erectus, 92

Homo habilis, 76
Homo sapiens, 26, 135
Hopkins, Phil, 224–25
"Hotel California," 18
Houdini, Harry, 39
Howard Hughes Medical Institute, 121,
      307
Hoyle, Fred, 153, 252, 281
Hubble, Edwin, 181–82, 250, 251, 254–
      55, 261–62, 319, 321
Hubble's law, 250, 252, 253
Hubble Space Telescope, 181–82, 191,
      232, 250
Huesmann, L. Rowell, 24
Hulse, Russell, 236
humanism, 200
Humanity+, 200
humans:
   bipedalism developed in, 76
   divergence of chimpanzees and,
      76–77, 95, 97, 101
   enhancing mental and physical capa-
      bilities of, 199–202, 200, 201
   evolution of, 26, 76–78, 75, 91–92,
      94–109
   interaction of bacteria and, 97
   large brain of, 76, 77, 135
   life span of, 25–26, 29, 32–33, 42–50
   migration of, 44, 76, 77, 95
   modifying of, 104–6
   potential future life of, 12, 101, 152,
      165, 195–98, 239–43, 280–90
   rapid ascent to culture and civilization
      by, 77, 135
   simulation of, 286–88, 288
   social cooperation of, 76–77
   species barrier between animals and,
      71
   survival and future of, 65–91, 106,
      137–65, 199–202, 280–90
   threats to, 12, 81–89, 280–87
   tools and technology developed by, 76,
      100, 101, 140
   see also death; hunter-gatherers; spe-
      cific human species
Human Touch, The (Frayn), 276, 324
Hume, David, 25

hunter-gatherers, 44, 76, 80, 100
Hunter Program, 88
hydrogen, 125, 188, *188*, 189, 191, 192,
    213, 215, 219, 234, 269, 273
hydrogen monoxide, 165
hydrogen sulfide, 118, 120, 152, 242
hydrothermal vents, 117, 118–19, *119*
hypernovae, 178, 180, 296
hypothermia, 61–62, 242

IBM Advanced Computing Systems
    Division, 238
ice, 114, 168–69
    glacial, 112, 129–30, 133, *143, 187,*
        311
    melting of, 129, 189
ice ages, 181, 185–87
immortality, 52, 55, 102–6, *103,* 285–
    86, 296
immune system, 88, 100
Industrial Revolution, 26
infant mortality, 43, 45–46
infection, 42, 46, *47,* 48, 66
influenza, 46, 153, 308
insects, 31, 43, 65, 78, 80, 91, 93, 120,
    205–6
Institute for Genomics Research, 117
insulin, 103
insurance companies, 42, *51,* 152
intelligence, *37,* 101
    artificial, *201,* 202
    extraterrestrial, 106–8, *107*
    see also life, complex, intelligent
intelligent design, 114, 277
Intergovernmental Panel on Climate
    Change, 81–83, *82,* 182
International Space Station, 184, 242
interstellar gas clouds, 281
Io, 168
Irish Republican Army (IRA), 22
iron, 121, 124, 139, 175, 215, 238, 273
    radioactive, 176, *176*
iron sulfide, 118
"iron-sulfur world" hypothesis, 118
Islam, 38
Ivanov, Ilya, 71
Ivins, Bruce, 87

Jainism, 39
Jansen, Karl, 40, 303
Jansky, Karl, 222
Japan, 32, 86, 111, 268
Japanese Encephalitis virus, 99
Jefferson, Thomas, 142
jellyfish, 55, 62, 83, 90–91
Jesus of Nazareth, 28, 141
jet propulsion, 149
Jet Propulsion Lab (JPL), 111
Joint Quantum Institute, 195
Jones, Dick, 59
Jones, Steve, 100–101
Joy, Bill, 105, 309
Judaism, 38
Jupiter, 136, 144, 156, 162, 163, 168,
    169, 198–99, 208, 232, 240,
    312–13

Kafka, Franz, 21
Kaltenegger, Lisa, 131–33
Kansas, University of, 179
Kant, Immanuel, 250
karma, 39
Kasting, Jim, 188
Kearl, Mike, 23
Keck 10–meter telescope, 222–23,
    317
Kelvin, Lord, 153
KGB, 87
kidney disease, 22, 80
Kirkaldy, G. W., 93
Knapp, Michelle, 142
Koch, Robert, 45
Korycansky, Don, 198–99
K-T extinction, 74, 78, 174
Kübler-Ross, Elisabeth, 39–40
Kubrick, Stanley, 221
kuru, 66–68
Kurzweil, Ray, 103–5, 109, 199, 284,
    308–9
Kyoto Protocol, 182

lactase, 101
Lahn, Bruce, 101
Landau, Lev, 288
Large Hadron Collider, 177–78, 241

Large Synoptic Survey Telescope, 147, 148
Largo, Michael, 23–25
Laser Interferometer Gravitational-Wave Observatory (LIGO), 235, 320
Lassa fever, 99
Laughlin, Greg, 198–99, 233
laughter, 65, 66
Lawrence, D. H., 21
Lawrence Berkeley Lab, 261
Lazcano, Antonio, 117
Lederberg, Joshua, 98
Leibniz, Gottfried, 277
leisure, 32, 100
Lemaître, Georges, 252
life:
    complex, intelligent, 106–8, 107, 109, 159–65, 189, 244, 246, 247, 271–72, 275–77, 283, 290
    control of matter by, 276
    dormant, 31, 153
    dreams and aspirations in, 20, 26
    essential ingredients of, 62–64, 106, 117, 160, 229
    extraterrestrial, 123, 133, 153–56, 161–62, 245–47
    human achievement in, 20–21, 26
    oldest forms of, 26–28
    origin and evolution of, 26, 114–18, 131–33, 132, 153, 158, 247, 313–14
    "primordial soup" idea of, 116, 117
    rebirth, or resurrection of, 34, 38–39
Life and Death of Planet Earth, The (Ward and Brownlee), 189
life expectancy, 25–26, 42–50, 296
    of animals, 28–30, 29, 43
    at birth, 42–43, 45, 46
    ethnicity and, 32, 48
    gains in, 42–43, 44–46, 45, 47, 52
    impact of wealth and poverty on, 43, 48
    scientific study of, 42
"life in a bottle" (Miller-Urey) experiment, 116
lifestyle, 49, 129
light, 125, 290
    speed of, 112, 195, 233, 241, 242, 256–57, 256, 259
    ultraviolet (UV), 133, 152, 153, 155
    see also sunlight
lightning, 181
Lincoln, Abraham, 23
Linde, Andrei, 279–80
Linnaeus, Carl, 68, 93, 305
Lion King, The, 22
lions, 28, 67, 71, 91
Little Ice Age, 181, 185–86
Liverpool, University of, 29, 30
Lloyd, Seth, 284, 286, 324
Loeb, Abi, 221
London, 40, 41–42, 72, 110, 113, 139, 143, 211
London School of Economics, 101
longevity, 27, 55, 57, 77, 81, 102–3, 161, 205–6, 240
Lord of the Flies (Golding), 126
Lovelock, James, 110–12, 126–29, 130–31, 311
Lucretius, 25
Lunar and Planetary Lab, 169
Lunar Prospector mission, 145
Lyme disease, 83

MacArthur "genius" Fellowships, 222
McAuliffe, Christa, 19
Macbeth (Shakespeare), 23
McKay, Chris, 119–20, 196–97, 310, 316
mad cow disease (bovine spongiform encephalopathy), 66
magnesium, 125, 215
magnetic resonance imaging (MRI), 17, 40
malaria, 44, 83, 101
mammals, 28, 29, 53, 92, 134, 161
    diversification and spread of, 135
    evolution of, 75, 75, 91, 95, 98
    extinction of, 135, 140
    hybrid, 71
    large, 30, 91, 95
    longevity of, 240
    see also specific mammals
mammoths, 140
Manicouagan crater, 143

Marburg fever, 87
Margulis, Lynn, 127–28
marine life, 27–28, 176, *176*
Mars, 111, 136, 137, 154–58, 163, 166–
    68, *167*, 190, 199
    arid and sterile atmosphere of, 131,
        196
    colonizing of, 195–97, *196*, 238, 316
    death of, 191
    gravity of, 154, 157, 158
    rocks ejected from, 154, 157
    testing for life on, 111, 119–20, 123,
        130–31, 155, 166, 309, 310
    travel from Earth to, 154, 158, 195–
        97, *196*
Mars Global Surveyor, 132, 142
Mars Reconnaissance Orbiter, 166
Mars Society, 196
masks, 18–19, *20,* 301
Massachusetts Institute of Technology
    (MIT), 284
mass media, 21, *24,* 177
materialism, 38–39, 303
maternal mortality, 45, 46, 50
mathematics, 41–42, 50, 51, *274, 278,*
    280, 320
Matrix, The, 287, 303
matter, 267–68, 273–74, 290
    control of, 276
    conventional theory of, *278*
    decay of, 268
    energy converted from, 274
    see also dark matter
Maunder minimum, 186
Maya, 89, 204–5
mayfly, 31, 205–6, 290
Mayr, Ernst, 68
medicine, 43–45, 58
    modern, 44–45, 47, 239
    oral tradition in, 43
    primitive remedies in, 43–44
mediocrity principle, 160, 165, 275
Meditations on First Philosophy (Des-
    cartes), 36
Mediterranean diet, *33,* 48
Mediterranean Sea, 27
Melosh, Jay, 145–46, 156

Melott, Adrian, 179, 314
memories, 33, 35, 58
    false, 40
    "folk," 142
    forming of, 18
Menudo, 34
Mercury, 168, 184, 190
metabolism, 57, 296
    of animals, 29–31, *30,* 55
    slowing of, 61–62, 242, 281
"Metaphysical Meditations" (Descartes),
    37
Meteor, 145
meteorites, 137–43, 154, *154, 155*
meteors, 113, 123, 157, 203, 290, 312
    see also asteroids; comets; meteorites
meteor showers, 141–42
methane (CH_4), 64, 121, 129, 130, 132,
    133
mice, 28, 29, 53, 83, 95, 105, 115, 242
Michigan, University of, *24,* 232, 264
microbes, 113, 120–24, 130, 133, 151,
    247, 306
    adaptation of, 152, 157–58
    unique ecosystems of, 96–98, 121–23
microbiology, 120, 127–28
microcephalin gene, 77, 101
microevolution, 74
microorganisms, 116, 123
Middle Ages, 21–22, 43, 50, 86
midwives, 45
Milanković, Milutin, 187, *187,* 296, 315
Milkomeda, 221, 223–26, 264, 266,
    268, 270, 290, 296
Milky Way, 12, 106–9, *107,* 112, 136,
    156, 159, 160, 175–80, 203–37,
    207, 250–52, 263, 279, 290,
    296, 317–18
    aging of, 227–37, 266, 319–21
    black hole in, 222–26, *224, 226,* 230,
        236, 237
    formation and evolution of, 211–37,
        214, 217, 220
    gravity of, 175, 211, 219, *220,* 228,
        232–37, *235*
    looking for habitable worlds in, 158–
        72, 195–97, *196,* 245–46

Milky Way (*continued*)
  mergers of galaxies in, 214, 218–21,
    220, 223–26, 229, 231, 235,
    237, 267, 318
  myths and legends of, 203–5
  place of Earth in, 211
  size, mass, and stellar content of,
    206–7, 209–11
Millennium Ecosystem Assessment,
  78–79
Millennium Simulation, *214,* 317–18
Miller, Stanley, 116, 117, 310
Miller-Urey ("life in a bottle") experiment, 116
minds, 202
  bodies vs., 36–38, *37*
  brain vs., 18, 36, *37*
  consciousness of, *37*
  demented states of, 52, 89
  thought processes of, *37*, 281, 284,
    303
minerals, 114, 122, 124, 127
Minsky, Marvin, 200
Miocene era, 76
mitochondria, 61
Mohave Desert, 27, 193
molecules, 11, 52–53, 62, 116, 118,
  273, 304
monasteries, 50
monkeys, 73, 75–76, *75*, 90, 94
Monroe, Christopher, 195
Moon, 145, 168, 199, 208
  craters on, 136, 142
  creation of, 114, 137–38
  full, 211, 289
  NASA missions to, 80, 81, 145, 190,
    243
moons, 56, *154*, 160, 244, 313
  habitability of, 168–70, 171–72, 240
Moore, Demi, 35
Moore's law, 246, 286
Moravec, Hans, 105, 283
morphology, 67–69, *68, 69*
mortality, law of, 42, 51, *51*
Mount Wilson Observatory, 181–82,
  250
M-theory, 282
mules, 71

Muller, Rich, 173–74
multiple sclerosis, 88
multiverse, 279–80, 282, 296, 324
murder, 24, 35, 49, 50
music, 30, 34, 36, 58, 221
mustard gas, 86
myelin, 88

nanotechnology, 104–6, 152, *201*, 298
Napoleon I, Emperor of France, 61
narcissism, 59
National Academy of Sciences, 222
National Aeronautics and Space Administration (NASA), 64, 144–45,
  147–49, *154*, 169, 192–93, 196,
  198, 240, 242
  lunar programs of, 80, 81, 145, 190,
    243
  solar programs of, 184–85
  *see also specific programs and
    spacecraft*
National Geographic Society, 95
National Public Radio, 50
Native Americans, 98
natural disasters, 49, 78, 94–95, *96*
Natural History Museum, 93
natural selection, 55, 74, 77, 100–102,
  118, 128, 135, 152, 240
nature, 39, 51, 62, 74, 276
Nature, 126
Neanderthal man, 43, 92, 102
near-death experience, 39–40
  imagery associated with, 17–18, 40
nebulae, 175, 250
nematodes, 92, 97
Nemesis, 173–74
neon, 125, 215
Neptune, 168, 169, 172
nervous system, 87
neurons, 59, 61, 88, 202
neutrinos, 175, 270
neutrons, 268, 273, 284, 322
New Horizons mission, 64
Newton, Isaac, *20*, 223, 251
Nibiru, 89
Nicholson, Jack, 51
nickel, 139
Nicomachean Ethics (Aristotle), 25

Nine Crazy Ideas in Science: A Few Might Even Be True (Erlich), 174
911 emergency, 60
99942 Apophis asteroid, 147–48
nitrogen, 125, 166, 169, 175, 197
Norman, Harold, 22–23
Northup, Diana, 120, 310
Nostradamus, 89
Notre Dame University, 39
Nova, 88
nuclear energy:
  chain reaction of, 127, 227
  fusion of, 227, 233, 318
  release of, 84, 106, 139, 145, 175
nuclear weapons, 84–87, 89, 106, 139, 149, 178, 269, 281, 290
nuclear winter, 85–86, 297
nucleic acids, 66–67, 117, 123–24

obesity, 100
Ocean Drilling Program, 123
oceans, 63, *63,* 112, 114, 116, 125–27
  circulation of, 152, 187
  condensation of, 150–51, 157
  life in, *68,* 77, 114, 125–26, 133, 134–35
  microbial biomass in, 96, 130
  oxygenation of, 152
  salinity of, 126–27
  sediment in, 126, 127, 129–30, 152, 176, 189
  warming of, 82, 83, 189
Odyssey, The (Homer), 102
oil, 125
Oligocene epoch, 67
Omega Point, 276
O'Neill, Eugene, 21
Oort Cloud, 172–75, *173,* 297, 314
Oparin, Alexander, 116
Ophiuchus constellation, 174
Ordovician extinction, 179
organisms, 26–27, 92, 106, 126, 128
  interbreeding of, 70–72
  recycling of, 62
  single-celled, 78
organs, 60, 104
Origin of Species, The (Darwin), 116
Orion constellation, 174, 211, 216

ouroboros, 254
Our Town (Wilder), 21
out-of-body experiences, 40
Oxford University, 202, 286
oxidation, 52–53
oxygen, 26, 124–25, 126, 130–33, *132,* 165, 175, 197, 228, 242
  absence or loss of, 18, 60–62, 116, 122
  infusion of, 61, 62
ozone cover, 86, *132,* 152, 176, 184

paganism, 44
paleontology, 92
Paleoproterozoic era, 130
Paleozoic era, 78
Pangea, 150
Panoramic Survey Telescope and Rapid Response System, 147
panspermia, 153–56, 297
paralysis, 88
parasites, 43, 44, 87, 97
Parkes, John, 123
Partial Test Ban Treaty of 1963, 84
particles, 56, 137, 211, 265–66, 297
Pasteur, Louis, 45, 115–16, 153, 309–10
Paul, Gregory, 106
Pennsylvania, University of, Center for Resuscitation Science at, 61
Pennsylvania State University, 188
Perlmutter, Saul, 261–62
Permian extinction, 77–78, 150, 151–52
personal identity, 34–35, 58, 307
pesticides, 80
phenotypes, 73–74, 297
Phobos-Grunt spacecraft, 156
phosphorus, 130
Photom M3 rocket, *155*
photons, 57, *57,* 184, 195, 254, 256–57, 259, 264–65, 268, *269,* 270, 275
photosynthesis, 56–57, 125, 130, 133, 152, 158, 189
physics, 56, 233, 254, 268, 279
  debate over fine-tuning in, 273–75, *274, 277, 279,* 294
  particle, 265–66
Picard horn, 259
Pilachowski, Caty, 190

Pilgrims, 27
Pindar, 19
pineal gland, 37
Pink Floyd, 191, 232
plague, 44, 86, 87
Plait, Phil, 179–80
Planck satellite, 259, 266, 282
Planetary Defense Conference of 2004,
    149
Planetary Society, 156
planets, 26, 56, 64
    dwarf, 168, 195
    formation of, 137, 234
    habitable, 12, 106–8, 107, 132, 160,
        163–68, 167, 310, 314
    magnetic fields of, 184
    reengineering of, 195–97, 196
    searching for life on, 130–32, 132
    terrestrial, 166–68, 167, 171, 208
    uninhabitable, 107
    see also exoplanets; specific planets
plants, 27–28, 72–73, 133, 125, 189
    asexual reproduction of, 70
    clonal colonies of, 27, 292
    flowering, 70
    hybrid, 67, 71
    oldest living, 27, 55
    species of, 67–68, 69, 78, 112
plate tectonics, 63, 130, 136, 159, 167
Plath, Sylvia, 21
Plato, 25, 36
Plutarch, 34
Pluto, 64
plutonium, 81, 84, 89
pneumonia, 46, 88
poetry, 21, 102–3, 126
polymerase chain reaction, 123
polymers, 117
polyploidy, 70
polyps, 55
Popov, Sergei, 88, 308
Popular Mechanics, 238
populations, 75–78
    bottlenecks of, 94–95, 96, 297
    connecting of, 100
    divergence of, 74, 76–77, 101

growth of, 101
    isolation of, 75, 77, 100
    urbanization of, 113
power system, 184
predators, 29, 42, 55, 76–77, 123
pregnancy, 59–60
Preskill, John, 285
primates, 31, 135, 159, 306–7
    evolution of, 74–77, 75
Princeton University, 160, 165, 236, 282
prions, 66, 87, 89, 96
Proceedings of the National Academy of
    Sciences, 86
Project Bonfire, 88
prokaryotes, 117, 123, 128, 132, 297,
    310
proteins, 62, 66, 73, 88, 89, 116, 117
protons, 268, 269, 275, 322
protozoans, 26
Proxima Centauri, 171, 174
Prusiner, Stanley, 66
Ptolemy, 208
public health, 24, 81
pulsars, 176, 218, 236
Pynchon, Thomas, 21, 143
pyruvate, 118, 119

Quake, Stephen, 121, 310–11
quantum mechanics, 278, 284
quantum states, 195, 248, 253, 254,
    269, 276, 279
quarks, 38, 278
quasars, 224–26, 229, 230, 257, 297,
    322

Rabelais, François, 21
rabies, 99
radiation, 57, 121, 146, 149, 153, 180,
    183, 213, 218, 259
    electromagnetic, 165, 320
    gravitational, 236
    Hawking, 284–86, 285, 295, 323
    microwave, 26, 253, 257–58, 264,
        322
    ultraviolet (UV), 133, 152, 153, 155,
        179

Radical Evolution (Garreau), 104
radioactive isotopes, 175, *187*
radio astronomy, 210
radio transmission, 109, 222, 231
rain, 136–37, 158
rainforests, 27, 65, 112
Rare Cat Hypothesis, 164
Rare Earth (Ward and Brownlee), 159
Rare Earth hypothesis, 159–65, 297
rats, 83, 91, 95
Reaction Engines Limited, 194
redwood trees, 67, 70, 134
Rees, Martin, 279–80, 324
reincarnation, 39
Reiss, Adam, 261–62, 267
relativity theory, 236, 250–51, 262–63,
    278
religious faith, 25, 302
    belief in life after death and, 23,
        33–34, 36, 38–39
    monotheistic, 34, 36, 38–39
    nontheistic, 39
    reason vs., 39
Renaissance, 26, 43, 200
RepRap project, 244
reproduction, 26–27, 53, 70, 100
reptiles, 28, 29, 49, 90, 91, 134, 135
Republic, The (Plato), 25
respiration, 40, 197
respiratory disease, 48, 98–99
retrotransposons, 75
Return of the Jedi, 169
Reynolds, Patti, 17–18, 39
Richter scale, 140, 146
RNA, 66–67, 72, 117–18, 298, 306
"RNA world" hypothesis, 117–18, 298
Robinson, Kim Stanley, 195
robots, 104, 105, 243–46
rockets, 192, 194, 197, 198, 240
rockfish, 27–28, 55
rocks, 113–14, 116, 120, 122, 124–25,
    127, 138–40, 142, 150–51, 152,
    155–56, 321
Rodale, Jerome Irving, 23
rodents, 90, 91, 95
Roman Empire, 21, 45, 50, 67, 86

roses, 67, 93
Roxanne, 51
Royal Society, *20,* 42
Rukeyser, Muriel, 11
Russia, 48–49, 85, 122
Russian Space Agency, 192, 193
Rutan, Burt, 193, 194

Saberhagen, Fred, 245
Sagan, Carl, 25, 85, 131, 195–96, 321
Sagittarius constellation, 206, 210, 222,
    224
Saint-Exupéry, Antoine de, 194
San Diego Supercomputer Center,
    218–19
sarin, 86, 87
SARS virus, 99, 153
satellites, 148, 174, 178, 184, 259, 320
Saturn, 168, 198–99, 232
Scaled Composites, 193
Schliemann, Heinrich, 19
Schmidt, Brian, 261–62
Schneider, Gregor, 22
Schweickart, Rusty, 149
science, 44, 142, 145
    extending fact to speculation in,
        11–12
    historical vs. experimental, 116
    myths of, 11
    organization and understanding
        through, 11
Science, 98, 307, 317
science fiction, 30, 89, 104, 145, 152,
    169, 195, 240–41, 281
scrapie, 66
sea core samples, 123
sea grass, 27
Search for Extraterrestrial Intelligence
    (SETI) Institute, 170, 245–46,
    321
Second Coming, The (Yeats), 267
self-awareness, 37, 108, 249
self-replicating space probes (bots), 244,
    246, 298
Sentry automated collision monitoring
    system, 144, 147

sex, 26–27, *69*, 70–72, 75, 100, 101, 281

Sexton, Anne, 21

sexual dimorphism, 69–70, *69*

Shakespeare, William, 27, 113, 211–12, 218, 229

Shapley, Harlow, 210

"Shine On, You Crazy Diamond," 191, 232

Shoemaker, Eugene, 144–45

Showman, Adam, 169, 313

Silent Spring (Carson), 79–80, 111

silicon, 175, 215, 284

simulation hypothesis, 272, 298

Sirius, R. U. (Ken Goffman), 200

Sirius A, *171*

Skoll Foundation, 99

sleep, 51, 281

Sleep, Norman, 114, 150–51

smallpox, 46, 87, 88, 98, 105

Snowball Earth, 129–30, 133, 150, 298

Society of Actuaries, 51

Socrates, 38–39

soil, 27, 69, 93, 111, 127, 140

Solar and Heliospheric Observatory (SOHO), 184

Solar System, 12, 112, 124, 131, 135, 136, 141, 154, 156, 172, 190, 197, 198, 263

  colonizing of, 157–80, 195–98, *196*, 238–45, *243*, 299, 312–14

  demise of life in, 64, 108, 175–80, *176*

  formation of, 26, 157, 217–18

  inner, 141, 147

  planets outside of, 131, 133, 159, 160, 162–63, *164*, 169–70, 294, 312–13

  threats to, 172–75

  see also Earth; planets; stars; Sun

solar wind, 184

soul, 34, 36, 37, *37*

Soviet Union, 84, 87–88, 178

space, 57, 63–64, 177

  biological evolution in, 106–9

  burial in, 64

Great Silence in, 246–47, 295

interstellar debris in, 136–44, *138*, 153, 157, 210

shape of, 253, 257–59

vacuum of, 56, 155, 248, 264, 275

Space Age, 26, 65

space colonies, 193–94, 239–47, *243*, 298

spacecraft, 80, 81, 109, 149, 192–94, 241

space elevators, 197

Space Island Group, 193

space probes, 131, *150*, 198, 243–46

  self-replicating, 244, 246, 298

space programs, 192–94

  Chinese and Russian, 192, 193

  private, 193–94

  see also National Aeronautics and Space Administration

SpaceShip Two, 194

space-time, 11, 12, 64, 234, 248, 251, 252, 254, 257, 258, 264, 265, 277, 278, 279, 280, 283

species, 29–30, 67–75, 298

  adaptive behaviors of, 95

  "ageless," 27, 55–56

  cognitive function of, 108

  definitions of, 67–68, 72

  divergence of, 74, 76–77, 95, 97, 101

  evolution of, 11, 72–77, 90–94, 189

  extinction of, 71, 76, 77–79, *79*, 91, 92, 94, 95, 96, 135–36, 140, 159, 295

  hybrid, 67, 70–72, 76, 90

  identifying, naming, and cataloging of, 92–94

  interbreeding of, 67, 70–72

  mating of, 68, *69*, 70–72, 75, 101

  mimicry in, 69

  morphology of, 67–69, *68*, *69*

  new, 72, 74–75, 78, 92, 93

  recovery of, *95*, *96*

  social cooperation within, 76–77

sperm, 60, 70, 100, 120

Spetzler, Robert, 17–18

spiders, 69, 70, 93, 308

spirits, 40, 44
Spitalfields Mathematical Society, 41–42
squids, 90, 91
stalactites, 120
Stalin, Joseph, 71
Stalingrad, Battle of, 87
Stanford University, 114, 150, 279, 280
Stapledon, Olaf, 195, 281
Star Maker (Stapledon), 281
stars, 13, 26, 56, 63–64, 160
    artificial, 222
    in binary systems, 171, 172, 234, 236, 237, 262
    corpses of, 231–34, 237, 298
    death of, 64, 108, 175–80, 176, 189, 190, 213, 216, 217, 228–31, 267, 269, 281, 315
    dwarf, 170–72, 171, 174, 188, 211, 216, 228, 230–34, 232, 233, 236, 268–69, 292, 299, 315
    "falling," 141–42
    fixed patterns of, 146
    formation of, 64, 106, 108, 175, 213, 214, 216, 217, 219, 224–25, 226, 228, 229, 231, 233, 250, 259–60
    habitability of, 170–72, 171
    high-mass vs. low-mass, 161–62
    massive, 216, 217, 218, 227–28, 230, 319
    neutron, 216, 217, 231, 234, 269, 297, 314, 320
    orbits of, 218, 223, 224
    Sun-like, 107, 107, 108, 170, 171, 184, 210, 211, 230
    transmutation of elements in, 215–17, 229, 230, 232, 234, 249, 273
    see also galaxies; Milky Way; Sun; specific stars
Star Trek, 195
Steinhardt, Paul, 282
stem cells, 26, 106
Stewart, Ian, 165
Stone Age, 65
Storm P, 11, 238
string theory, 278–80, 278, 282, 324

stromatolites, 92, 113, 298
Stross, Charles, 241, 320
suicide, 20, 21, 22, 24, 49, 50, 87, 304
sulfur, 121
sulfur dioxide, 120
Sun, 12, 26, 57, 57, 113, 141, 154, 159, 172–74, 180, 181–92, 208, 255
    aging of, 182, 187–90, 188, 196, 198
    auroras from, 184
    climate change effected by, 181, 182, 183, 185–87
    coronal mass ejections from, 184, 185, 299
    death of, 187–92, 225
    formation of, 217–18, 234
    fusion reactions in, 188, 188, 190–91, 215
    gravitational influence of, 64, 135, 188, 263
    hypothetical stellar companion of, 173–74
    increased energy of, 114, 126
    interior of, 184–85
    life on Earth sustained by, 182, 185
    magnetic field of, 181, 184, 185, 186
    release of energy from, 114, 126, 183–85, 187, 191
    rising and setting of, 79, 112, 289
    total eclipses of, 185
sunlight, 56–57, 57, 114, 116, 122, 127, 129, 281
Sun Microsystems, 105
sunspots, 183–84, 208, 314–15
    agricultural production and, 183
    cycles of, 181, 184
    low numbers of, 186
Superfund, 80
supergalaxies, 214
Super-Kamiokande experiment, 268
supernatural experience, 37–38, 287
supernovae, 175–79, 176, 189, 216, 218, 219, 230, 234, 251, 261–62, 299, 317, 322–23
superstition, 44
supertaxa, 91
surgery, 17–18

suspended animation, 195, 242, 299
Swayze, Patrick, 35
Swift, Jonathan, 183, 237
Swiss Re, 152
Szostak, Jack, 117

taxonomy, 93
Taylor, Joe, 236
technology, 104–6, 135, 194–95, 199–
    201, 316
  dangers of, 105–6, 109
  human use of, 76, 100, 101, 140
  progress of, 238–39, 239, 243–45,
    246
  regulation and control of, 106
  see also computer technology;
    nanotechnology
tektites, 140
teleology, 277
teleportation, 194–95, 315–16
telescopes, 144, 146, 147, 148, 178,
    181–82, 186, 191, 192, 208–9,
    210, 222–23, 250, 255, 286,
    312, 320
telomeres, 54, 54, 55, 299
Tempest, The (Shakespeare), 27
Terminator, 245
terraforming, 195–98, 196, 245, 299
terrorism, 25, 49, 86–89, 144
testosterone, 120
thanatology, 23, 299
thermodynamics, 57, 281
  second law of, 270
Theseus, 34, 35, 38, 39
think tanks, 152, 182, 286
3200 Phaeton Comet, 141
This Is Spinal Tap, 26
Thomas, Dylan, 229
Thoreau, Henry David, 21
Through the Looking Glass (Carroll), 98,
    256
time, 11, 26, 160
  compression and distortion of, 213
  deep, 229, 268
  flow of, 56, 57
  see also space-time

Time Machine, The (Wells), 101.
"Time Without End: Physics and Biology
    in an Open World" (Dyson), 280
Tintoretto, 204
Tipler, Frank, 276
Titan, 168, 172, 191
Titanic, 270
TNT, 139, 140, 145, 269
Toba supervolcano, 94–95
Today Show, 32
Tolstoy, Leo, 21, 51
Tombaugh, Clyde, 64
Torino scale, 144
transhumanism (H+), 199–202, 200,
    201, 299
trees, 27, 134, 189, 205–6
trepanning, 44
Triassic period, 152
trilobites, 77, 93
Trinity College, 23
Triton, 168
Trojan War, 19
tsunamis, 199
tuberculosis, 44, 46, 239
Tularemia, 87
tundra biome, 112
Tunguska impact, 138, 139, 141
Turok, Neil, 282
Turritopsis nutricula, 55
turtles, 28, 55
Tutankhamen, 20
Tuthill, Peter, 180, 314
2001: A Space Odyssey, 221
Tyrannosaurs, 134
Tyson, Neil, 177

unidentified flying objects (UFOs),
    245
United Kingdom, 28, 32, 40–42, 45, 48
United Nations (UN), 78
United States, 36
  centenarians in, 32
  nuclear arsenal of, 84, 85
  waste generated in, 80–81
universal replicators, 244–45
universe:

age and history of, 26–28, 63, 106,
    213–15, 252, 256, 256, 261, 276
contents of, 261
counterfactual realities and, 273–75,
    293
death of, 202, 248–70, 321–23
growth and expansion of, 11, 182,
    213–15, 214, 249–52, 251, 255–
    57, 256, 259–65, 261, 281, 321
possibility of life in, 153–56
potential future paths for life in,
    65–91, 137–65, 199–202, 238–
    46, 272, 280–90
rhythm of, 56
role of intelligent observers in, 275–
    77, 276, 280
six most abundant elements in, 125
size of, 211, 255, 289
see also big bang
uranium, 80–81, 84, 89, 122, 122
Uranus, 163, 168, 169, 209, 313
Urey, Harold, 116
Uruk, First Dynasty of, 102

V-2 bombs, 143
vaccines, 87, 88
vacuums, 56, 155, 248, 264, 275, 278–79
Valentine, Stephen, 59
van Gogh, Vincent, 32
van Inwagen, Peter, 39
Venter, Craig, 117
Venus, 166–68, 167, 190, 195–96
Verne, Jules, 110, 122
Viking missions, 111, 120, 130, 309
Virgin Galactic, 194
Virgo Cluster, 264
viruses, 87, 88, 96, 98–99, 99, 151, 153
vitamins, 53, 62
volcanic eruptions, 63, 116, 125, 127,
    130, 142, 150, 151–52, 187,
    222, 228
von Braun, Wernher, 193
von Neumann, John, 244, 320

Wächtershäuser, Günther, 118
Walden (Thoreau), 21

Wall Street Journal, 51, 305
war:
    biological weapons used in, 86–88
    death in, 50, 85–86
    nuclear, 85–86
Ward, Peter, 91, 159, 160–61, 189
Washington, George, 34
Washington, University of, 35, 91, 159
waste, 80–81
water, 49, 62, 117, 132, 132, 159, 160
    boiling and freezing points of, 120
    pH conditions in, 120
    on planets, 166, 167, 169, 171, 197
    salinity of, 120, 126–27
Watson, Thomas, 238
weapons of mass destruction (WMD),
    88–89, 308
weather, 83, 94, 183
weight loss, 62
Weinberg, Steven, 277, 279–80, 324
Weismuller, Johnny, 31
Wells, H. G., 101, 110
Wells, Spencer, 95
West Nile virus, 83, 99, 99
whales, 43, 108, 138, 230, 238, 290
What Does a Martian Look Like? (Cohen
    and Stewart), 165
Wheeler, John, 12, 276
Wheeler, Quentin, 93
When Worlds Collide, 145
White-Stevens, Robert, 80
Wickramasinghe, Chandra, 153
Wilder, Thornton, 21
Wilkinson Microwave Anisotropy Probe,
    253, 253, 258
Williams, Ted, 59
Williams, Tennessee, 21
Wired, 105
World Health Organization, 81
World Trade Center destruction, 25
World Transhumanist Association, 200,
    202
World War I, 86
World War II, 86, 87, 143
worms, 53, 70, 93, 97
WR 104, 180

X-rays, 222, 225

Y chromosomes, 94
Yeats, William Butler, 52, 267
yellow fever, 87
Y2K, 89
Yucca Mountain, 81

Zigas, Vincent, 65–66
Zircon, 114, *115*, 299
zoos, 28, 72, 83
Zubrin, Robert, 196–97